THE DECLINE OF THE GURU

THE DECLINE OF THE GURU
THE ACADEMIC PROFESSION IN THE THIRD WORLD

Edited by
Philip G. Altbach

palgrave
macmillan

First published 2003 by
PALGRAVE MACMILLAN™
175 Fifth Avenue, New York, N.Y. 10010 and
Houndmills, Basingstoke, Hampshire, England RG21 6XS
Companies and representatives throughout the world

PALGRAVE MACMILLAN is the global academic imprint of the Palgrave Macmillan division of St. Martin's Press, LLC and of Palgrave Macmillan Ltd. Macmillan® is a registered trademark in the United States, United Kingdom and other countries. Palgrave is a registered trademark in the European Union and other countries.

0–312–29591–X hardback
1–4039–6163–8 paperback

Library of Congress Cataloging-in-Publication Data

The decline of the guru: the academic profession in developing and middle income countries / Philip G. Altbach, editor.
      p. cm.
      Includes bibliographical references and index.
      ISBN 1–4039–6163–8 (pbk) – ISBN 0–312–29591–X (cloth)
      1. College teachers – Social conditions – Cross-cultural studies.
   2. College teachers – Selection and appointment – Cross-cultural studies. I. Title: Academic profession in developing and middle income countries. II. Altbach, Philip G.

LB1778 .D43 2002
378.1'2 – dc21   2002035260

A catalogue record for this book is available from the British Library.

Design by Newgen Imaging Systems (P) Ltd., Chennai, India.

First Palgrave Macmillan edition: May, 2003
10 9 8 7 6 5 4 3 2 1

Printed in the United States of America.

CONTENTS

PREFACE

This book has multiple antecedants. It stems from several initiatives relating to the changing academic workplace and academic appointments in North America and Europe. Professor Richard Chait of Harvard University directed a multidimensional project on academic appointments in the United States, with funding from the Pew Charitable Trusts. The Harvard Project on Faculty Appointments was concerned with analyzing and fostering discussions about the terms and conditions of academic appointments in the United States. With those goals in mind, the project organizers felt that a careful look at the situation in Europe would provide an important comparative perspective. Somewhat earlier, Professor Jürgen Enders, now at the Centre for Higher Education Policy Studies at the University of Twente in the Netherlands, coordinated an inquiry sponsored by the European Union concerning "Employment and Working Conditions of Academic Staff in Europe." With the help of Professor Enders and the sponsorship of HPFA, I organized a group of scholars who prepared essays on the changing academic profession in Europe and the United States. That effort resulted in a special theme issue of *Higher Education* (vol. 40, no. 1–2, 2001) and a book, *The Changing Academic Workplace: Comparative Perspectives,* published in 2000 by the Center for International Higher Education at Boston College.

The academic profession is undergoing dramatic changes in developing countries, and this book reflects our effort to shed light on these key parts of the world. With the assistance of the Ford Foundation, we commissioned research studies in developing and middle-income countries and met at the Villa Serbelloni, the Rockefeller Foundation's conference center in Bellagio, Italy, to discuss the research. Based on the conference, the authors revised their work and this book is a result of this fruitful collaboration.

Three of the essays prepared for our conference are not included in this book—these essays focus on Central and Eastern Europe, and it was decided that this volume would focus exclusively on the Third World. We are indebted to Marek Kwiek, Snejana Slancheva, and Anna

Smolentseva for their participation in our project. These essays, dealing with Poland, Bulgaria, and Russia, are included in a volume published by the Center for International Education at Boston College, and will be published in a theme issue of *Higher Education* in 2003.

I am indebted especially to the Ford Foundation and the Rockefeller Foundation for their financial support of this project. Dr. Jorge Balan, program officer for higher education at the Ford Foundation, has been supportive of our work and has provided central intellectual guidance as well. The Rockefeller Foundation's Villa Serbelloni was an incomparable venue for our conference. Edith Hoshino edited the manuscripts. Salina Kopellas, Robin Matross Helms, and Laura Rumbley of the Center for International Higher Education assisted in the preparation of this book. This book is part of the research program of the CIHE. Ongoing support from the Ford Foundation and from Boston College is much appreciated.

Philip G. Altbach
Chestnut Hill, Massachusetts
November 2002

1

CENTERS AND PERIPHERIES IN THE ACADEMIC PROFESSION: THE SPECIAL CHALLENGES OF DEVELOPING COUNTRIES

Philip G. Altbach

The academic profession worldwide is united by its commitment to teaching and the creation and transmission of knowledge. Yet, as pointed out by Burton Clark, it is also composed of "small worlds, different worlds" divided by discipline, role, and other factors (Clark, 1987). This chapter examines the conditions of the academic profession and workplace in developing countries. A growing proportion of the world's postsecondary students are found in developing countries, and the rate of expansion of higher education is greatest in this part of the world. By the mid-1990s, 44 million of the world's 80 million postsecondary students were in developing or middle-income countries—despite the fact that only 6 percent of the population in these countries have attained postsecondary degrees, while 26 percent in high-income nations have similar qualifications (Task Force on Higher Education and Society, 2000, pp. 111 and 115). Further, many developing countries are building up large and complex academic systems, including research universities. Yet very little is known about the professionals who are responsible for teaching and research in these universities.

What we do know about the conditions of the academic profession and of academic work in the developing world is not positive. Conditions of work and levels of remuneration are inadequate, involvement in institutional governance is often very limited, and the autonomy to build both an academic career and academic programs in the university is often constrained.

While some of these circumstances do exist in middle-income nations such as the countries of the former Soviet Union, this chapter is mainly concerned with developing countries—low per capita-income nations. There are major variations among the developing countries and, indeed, within the academic systems of these countries. The larger countries, such as India and China, have some universities and specialized postsecondary institutions with excellent facilities that operate at the highest international levels. However, overall, these higher education systems are not of a high caliber. Some of the smaller countries do have a number of academic institutions with high standards of teaching and research. Some countries have given a higher priority to higher education than others, and some countries have higher literacy rates or per capita income than others. Cuba, for example, has a high literacy rate and educational attainment although its per capita income is low. In Latin America, Mexico, Chile, Brazil, and Argentina are no longer classified as developing countries and have relatively high income levels and large and sophisticated academic systems. Yet some of the conditions common to poorer nations are found throughout Latin America. As with most comparative analysis, the generalizations presented here will not perfectly fit all of the countries or higher education systems discussed in the chapter.

WORLDWIDE TRENDS

Many of the conditions affecting the academic profession in developing countries are central realities worldwide. For example, G. R. Evans (2002) points out that the British academic profession has been drawn away from its traditional values and that in many ways this has weakened universities. As she points out, these trends are observable everywhere. The central realities of higher education in the 21st century—massification, accountability, privatization, and marketization—shape universities everywhere, and those who work at them, to differing degrees. Massification has led, among other things, to an expanded academic profession and an academic community that is increasingly unrecognizable. Accountability has limited the traditional autonomy of the profession, more tightly regulating academic work and eroding one of the major attractions of the academic profession. Privatization has, in some contexts, placed pressure on academics to generate income for themselves and for the university through consulting and other nonteaching activities. Marketization has forced academics to be more cognizant of student curricular interests and opportunities for entrepreneurial activities. The sad fact in the era of mass higher education is that the conditions of academic work have deteriorated everywhere.

While the current realities are not necessarily detrimental for either the profession or for higher education, they do constitute a major shift in the nature of academic institutions and academic work. The changes have implications for the career structure of the professoriate, choices for research and teaching, the relationship of academic staff to administration, and the participation of academics in the governance of institutions, to mention a few factors. In the industrialized nations, a segment of the increasingly differentiated academic systems have managed, so far at least, to retain the ability to engage in high-level teaching and research and to protect the central values of the university. While the problematic trends described earlier impact academics everywhere, the severity may be especially great in developing countries, where the traditional roles of the professoriate are often less well established, the financial and other resources less adequate, and the pressures greater.

CENTERS, PERIPHERIES, AND DEPENDENCY

The professoriate in the developing countries of the South is a profession on the periphery (Altbach, 1998). Research is, with few exceptions, undertaken at the major universities in the industrialized countries of the North, and the patterns of academic work in these institutions set the standard everywhere. The academic world is itself hierarchical, and research universities in the industrialized countries are at the center of an international knowledge system (Shils, 1972). These research universities set the patterns, produce the research, and control the key international journals and other means of communication. Academics in teaching-oriented universities in the North are peripheral to those at the major research universities in the North. Those in developing countries are also peripheral to the international centers.

The academic systems of the developing countries are, without exception, imported from the North. Indeed, all contemporary universities are based on the medieval University of Paris model with the exception of the al-Azhar University, in Cairo. In part, the European model was imposed by the colonial powers, but even in Ethiopia, Thailand, and Japan, where foreign academic patterns were not imposed, European models prevailed over existing indigenous academic traditions. Following independence, when developing countries had the chance to change the nature of the university, none chose to do so. Indeed, in many cases, even the language of the colonial power was retained for instruction and research. The major European languages remain dominant in many developing countries; for example, English and French are still entrenched in Africa. Indeed, no African languages (unless one chooses to count Arabic and Afrikaans as African languages) are used in African universities. Although India uses many of its regional languages for instruction, English remains important and preferred by many. The European and increasingly the American academic model—based on departments, competition among academic staff, institutional hierarchy, and specific definitions of science and scholarship—continues to prevail throughout the Third World.

There are many aspects of the peripherality of the academic profession in developing countries. Language is one element. In the 21st century, English is the main language for academic communication—in journals and Internet networks, as well as at international meetings. Other major Western languages such as French and, to a lesser extent, German and Spanish are also widely used. Other languages may be used for teaching and perhaps local publication, but have little international relevance for scientific research. A significant number of developing countries use English or French for instruction, creating global links—but also weakening the connection to local cultures and realities.

The long-established academic communities of the North are larger and wealthier than their counterparts in the South. Their resources permit them to maintain leadership in all aspects of academic work in ways that are difficult, if not impossible, for universities elsewhere to emulate. Universities in the North also have close relationships with multinational corporations and other users of research, thus giving them further sources of funding and outlets for research and other academic work. This combination of wealth, resources, and position cements the centrality of the universities of the North.

The current status of the academic communities at the center is, at least for the immediate future, an immutable reality of the world knowledge system. Universities in developing countries, and their academic communities, must function in the unequal world of centers and peripheries. Peripherality does not mean that academics in the developing world cannot do creative scientific or intellectual work, or that they are forever relegated to a subordinate status in academe. It does mean that they will seldom be at the frontiers of world science and will not share in the control over the main levers of academic power worldwide.

Related to peripherality is dependency. Third World academics often perceive themselves to be dependent on the main centers of knowledge and the world scientific networks. The vast inequality in wealth, size, and access to resources and institutional infrastructure contributes to dependency. The policies and practices of the academic systems in the North also play a role in the power imbalance. For example, scholarly journals select articles based on the interests as well as the methodological and scientific norms that prevail in the North, which often places Third World researchers at a disadvantage in getting their work published and recognized internationally. Funding for research, participation in international conferences and programs, and access to academic collaboration comes entirely from the North. The decision-making structures are based in the North and reflect the interests and concerns of the dominant academic communities. The situation is most extreme for Africa, where almost all research and funding for international linkages come from external sources— foreign governments, multilateral agencies such as the World Bank, philanthropic foundations, and others. African scholars and scientists are dependent on foreign funds and the particular priorities and programs of the funders for their research (Teferra and Altbach, 2002).

The fact that academics in developing countries function in a world of peripherality and, to varying degrees, dependency, is central to understanding the nature of academic work and the role of universities. While the professoriate everywhere is increasingly part of a global academic community, the wealthier and better-developed university systems of the North have more autonomy and resources with which to support independent teaching and research. Thus, while academe worldwide is increasingly affected by the power and influence of the largest academic systems, and especially those that use English, the developing countries are at the bottom in a world system of unequal academic relationships. By its nature, scholarly work in the 21st century is interdependent; the developing countries are importers of knowledge, and have little, if anything, to offer in return.

Having pointed out this context of inequality, the fact is that most academics work in the "small worlds" of their departments and universities, are mostly if not exclusively involved in teaching, and are thus unaffected in their daily lives by the trends of international scholarship. Even now, academic work for the most part takes place in a national context at universities that, while affected by global trends, nonetheless work on a day-to-day level in their own context.

THE DOMINATION OF EXTERNAL VALUES

The universities of the developing world are closely tied to the Northern-dominated system. Not only is the institutional model and often the language of instruction adopted, but many of the norms and values of the academic profession have been as well. The Third World looks to the North for validation of academic quality and respectability. For example, academics are expected to publish in Northern academic journals in their disciplines. Promotion often depends on such publication. Even where local scholarly publications exist, they are not respected. While it is understandable that small and relatively new academic systems may wish to have external validation of the work of their scholars and scientists, such reliance has implications for the professoriate. For example, internationally circulated journals are often highly competitive, and journal editors may not place much value on research topics relevant in developing countries. Moreover, it is always more difficult for authors to write in a language that is not their own. Journal editors, for their part, must be guided by the methodological and topical predilections of their immediate colleagues and are as a result less interested in work done by Third World authors. These authors are also at a distinct disadvantage because they do not have access to the library and laboratory facilities available at the major universities of the North.

In many ways, Third World academic systems rely on the North to validate their academic work. China and other developing countries measure the research productivity of academics in part by relying on the Science Citation Index and, to a lesser extent, the Social Science Citation Index. These measurements of the impact of scholarly work count citations in a group of internationally circulated journals. The number of journals covered is only a small proportion of those published, and almost all of them are edited and published in the North, thus systematically undercounting scientific work in the developing countries. Yet, these indexes are the only major sources available. Their influence emphasizes the importance of participation in international scientific networks, and undervalues scientific work carried out in developing countries that may be directly relevant to local needs and conditions. Even as Third World academics attempt to keep abreast of world science, they are at a distinct competitive disadvantage. The way in which the world of scientific publishing is organized discourages national and regional scientific communities from emerging in the Third World.

In most respects, academics in developing countries look to the North for both validation and models of higher education development and professional norms. While understandable and probably necessary for universities seeking to engage in research and teaching at the highest international levels, an overreliance on these external norms distorts academic development and introduces unrealistic expectations for institutions and for the academic profession.

THE IMPACT OF GLOBALIZATION

Global higher education developments have had a broad structural impact on systems everywhere (Scott, 1998). There are also elements of globalization that specifically affect the academic profession. The most visible aspect of globalization is the emergence of a worldwide market for academic talent, stimulated in part by the large numbers of students who study abroad. It must be emphasized that flows of foreign students and the international labor market for scholars and scientists are overwhelmingly a South-to-North phenomenon. Approximately 1.5 million students study outside the borders of their own countries—the vast majority of these students are from developing countries and their destinations are in the industrialized nations. The United States is the host country for 547,000 students. Western Europe, Australia, and Canada absorb most of the rest. There is only a tiny flow of students from North to South, although there is some South–South flow. A large majority of international students from developing countries study for advanced degrees—in contrast to patterns from the industrialized nations, where students tend to study for their first degree or spend just a semester or year abroad. A significant number of students who obtain their degrees abroad do not return home, and those who do return and join the academic profession bring the values and orientations of the country in which they studied back with them.

While foreign study has received considerable attention, its impact on the academic profession has not been analyzed. In many developing countries, academics with foreign degrees constitute a significant part of the professoriate. More important, these returnees are clustered at the top of the profession and dominate the research-oriented universities. They are the "power elite" of the academic community. This is the case for a number of reasons. Foreign academic degrees are valued not only because of the perceived quality of the training and the exposure to the best facilities and professors available but because foreign study is deemed to be more prestigious than staying at home for training. Scholars returning from abroad often wish to implant the values they absorbed during their studies to upgrade local standards, whether or not such replication is practical or desirable in local conditions. These academics follow the latest international academic developments and seek to maintain links with the country in which they studied, often importing scientific equipment as well as ideas. Conflicts between foreign-returned academics and their locally educated colleagues are common.

There is also an increasingly important flow of academic talent around the world. Again, the flow is almost exclusively from South to North. It takes many forms, including migration from one country to another on a permanent basis, stints as visiting scholars or postdoctoral fellows, or temporary work assignments abroad. Statistics are difficult to obtain, but some 80,000 visiting scholars were at American universities in 2000. We know that there is a large flow of academics from a number of African countries to North America and Europe— for example, more Ghanaian medical doctors are practicing outside of Ghana than at home. There is now a flow from sub-Saharan African nations to South Africa, while at the same time South African academics are taking jobs in the North.

What used to be called the "brain drain" is now a much more complex phenomenon. For academic and scientific personnel, settling in another country no longer means permanent emigration. In some cases, people from developing countries who take jobs in the North return home when attractive opportunities open up and the circumstances at home are appropriate in terms of living conditions, academic infrastructures, and the intellectual and political climate. As Taiwan and Korea developed in the 1960s and as both countries became stable democracies, academics and scientists who had settled abroad returned home to take jobs in universities. More common is the phenomenon of scientists and scholars from developing countries who have emigrated maintaining active relationships with their countries of origin (Choi, 1995). They serve as consultants, visiting professors, lecturers, or advisers to universities, governments, and sometimes companies in their countries of origin. In this way, they act as important links between centers and peripheries. They understand conditions at home and often retain a certain commitment to their home countries that expresses itself in academic cooperation and assistance.

In the 21st century, the diaspora of professors, scientists, and intellectuals from developing countries who study or live in the North is a significant factor in the academic culture of the developing world. Globalization makes this human flow possible. An international academic culture, the willingness of universities worldwide to accept students and in many cases faculty members from abroad, and immigration policies that permit migration all contribute to this diaspora. While the bulk of the flow is from South to North, there is also significant movement among the industrialized countries, with academics moving from countries with relatively low salaries and poor working conditions to those with greater resources. For example, large numbers of academics from the former Soviet Union have moved to Western Europe and North America in recent years. Small numbers have gone from the United Kingdom to the United States and Canada because of deteriorating salaries and working conditions in the United Kingdom. There has also been a modest South–South flow—Indians can be found teaching in a number of English-speaking African countries, and South Africa has attracted academics from other African countries. Egyptians and Palestinians staff universities in the Gulf and Saudi Arabia. The costs and benefits of this massive international migration are considerable—with most of the benefits accruing to the wealthier academic systems.

Information technology (IT) is also very much related to globalization and is beginning to affect universities and the academic profession in many ways. Two basic elements are of concern here—the use of IT for scientific communication worldwide and the use of IT for pedagogical purposes both through distance education and for improving instruction and learning in traditional universities. The information technology revolution is in its early stages and will increase its impact on higher education everywhere. For academics in developing countries, IT has so far had a considerable impact and will inevitably expand its role.

IT is a new phenomenon in much of the developing world—Africa, for example, has been connected to the Internet for just a few years, and even now many African academics have only sporadic access to it (Teferra, 2002). The issue of access is central. In the academic context of developing countries, many academic staff do not have their own computers and must rely on sharing with others. Personal e-mail accounts are by no means universal. Connectivity is sporadic and, in general, slow, as it still relies on inadequate and poorly maintained telephone systems in many places. Slow and sporadic access means that many of the sophisticated databases will not run well. Despite these serious problems, IT has provided many developing country academics with unprecedented access to current scientific information, to some extent making up for the inadequate libraries that exist in virtually all developing countries. Just as important, the Internet has permitted academics to communicate with colleagues worldwide, dramatically decreasing traditional isolation.

While IT has given access to knowledge on a scale hitherto unknown, it has in some ways increased the peripherality of developing country academics (Castells, 2000). Studies show that developing countries use information from the North, but contribute relatively little to the total flow of knowledge. They are, in sum, users of knowledge produced by others.

IT is beginning to come into its own as a means of delivering higher education through distance education courses and degree programs offered on the Internet. Developing countries are making use of distance education in this way—indeed, seven out of the ten largest distance education providers are located in developing countries—in countries such as Turkey, India, and China. With the exception of the few academics who have been involved with developing and delivering curriculum in these distance-based universities, few individuals in developing countries have had their teaching affected by IT. Even fewer have been able to use IT to enhance classroom-based teaching, as they lack the necessary facilities, equipment, knowledge, and funds. It is likely that these constraints will continue for some time and that IT use will be minimal in developing country universities. It is also the case that many of the IT "products" available worldwide originated in the industrialized countries, and are often available from media corporations. Designed for use in the North, they may not be relevant for developing countries. They may also be too expensive. The fledgling African Virtual University, for example, has had trouble finding "content" relevant for African countries. Creating courses and related content with a developing country perspective is both expensive and requires skills generally unavailable.

The impact of the Internet and IT on the academic profession is, in many ways, similar to the patterns of inequality described earlier in the chapter. Academics in developing countries are dependent for the technology, basic equipment, and content on outsiders. These technologies have helped developing country academics to keep abreast of scientific research, communicate with international colleagues, and participate in scientific debates on a more equal basis. However, they are still peripheral in many ways in the Internet-based knowledge system.

THE SHAPE OF THE PROFESSION

The professoriate is changing in many parts of the world—and the developing countries are not immune from these changes. In developing countries, a higher proportion of academics work on part-time contracts or are subject to irregular hiring practices. In many developing countries, a large part of the profession is composed of part-time staff who teach a few courses and do not have regular academic appointments or real links to the university. This is the norm at most

Latin American universities, where full-time permanent staff are a tiny proportion of the total academic labor force. In many countries, tenure is not guaranteed, and even full-time academics have little formal job protection—although, in fact, relatively few are actually fired. Clear guarantees of academic freedom or the assurance of a stable career are often missing.

There are curious contradictions in the nature of academic appointments. On the one hand, those hired in regular full-time positions are generally given de facto security of appointment, without much evaluation as to job performance, competence in teaching or research, or other attributes of a successful academic career. At the same time, while few are in fact removed from their academic posts, many academic systems do not offer a formal tenure system that protects academic freedom or inhibits interference by university authorities in the intellectual life of academic staff.

In many Latin American countries, the pattern of academic appointments includes periodic "contests" for academic posts, which require each professor to defend his or her position publicly and permits others to apply for the post. Often, contests do not occur due to the inability of university authorities to organize open competitions on a regular basis. In reality, few faculty are removed from the posts they already hold, but the possibility of removal remains. In many developing countries, the terms and conditions of academic appointments are not clearly spelled out, leaving considerable latitude for administrative or governmental interference in an academic career.

The requirements for academic appointment vary greatly in developing countries and are in general less rigorous than is the case in most industrialized nations. In the North, the standard requirement for an academic appointment includes holding a doctoral degree or the equivalent—the highest degree possible in the country. In Germany, Russia, and other countries following the German academic model, a second doctorate, the habilitation or its equivalent, is required for appointment to a full professorship.

Many of those who teach have earned only a bachelor's degree. It is probably the case that a majority or significant minority of academics in virtually every developing country hold just a bachelor's degree. Those in senior academic positions almost always have higher academic qualifications, but much of the academic labor force has modest qualifications for their jobs. A number of countries, including India and Brazil, have engaged in successful efforts to increase the qualifications of their academic staff by providing opportunities for study to those already in academic positions and increasing the minimum qualifications for appointments. The lack of qualifications has meant that academic upward mobility is limited for many junior staff. It also means, of course, that the level of expertise possessed by many teachers is quite modest, affecting the quality and depth of the instruction provided to many students.

It is unlikely that, on balance, the qualifications of academic staff will improve dramatically in the coming period. Continued expansion throughout the developing world means that large numbers of new teachers will be required, and selectivity will be minimal. The bulk of enrollment growth worldwide will be in developing countries—in India, for example, enrollments will almost double in the next two decades. The challenge of providing teachers to instruct these students will place severe strains on the limited capacities in most developing countries for advanced training in the universities.

The mixed qualifications of academic staff have resulted in a highly differentiated academic profession, with the small minority of well-qualified professors, many holding foreign doctoral degrees, at the top of the system, and the large majority of poorly qualified teachers at the bottom, with few possibilities for mobility. Missing is a successful middle rank of scholars. A wide gulf exists between the thin wedge of highly qualified personnel at the top of the system and the large, poor, and marginally qualified group of teachers at the bottom.

We have only limited knowledge of the socioeconomic backgrounds of the academic profession in developing countries, but some generalizations can be made. The involvement of women in the profession varies and is surprisingly high in some countries. In many Latin American nations and in South Asia, the proportion of women holding academic positions is high—often higher than in industrialized countries. As of 1993, more than a third of academics in three large Latin American countries (Brazil, Mexico, and Chile) were women (Boyer, Altbach, and Whitelaw, 1994). Only a few industrialized nations have reached that level. In developing countries, academics tend to come from well-educated, urban families, although the majority of the population remains largely uneducated and rural. Academics do not, however, come overwhelmingly from elite families, due in part to the fact that salaries are not high and chances for mobility are limited.

The academic profession in developing countries differs significantly from the professoriate in the North. There are more part-time staff. Full-time professors have less job security and are sometimes subject to insecure terms of appointment. They are not as well qualified, and they come from more modest backgrounds. While there have been efforts to upgrade academic skills in some developing countries, massification has meant that qualifications have not kept up with the need for more teachers in the classrooms of the Third World.

DIFFERENT REALITIES

While it is, of course, true that the basic roles of academics everywhere are similar—teaching, research, and service, in different proportions, are the central responsibilities—it is also true that in all countries, most academics are mainly teachers, with research and service a minor or negligible part of their work. It is also the case that academics world-wide have recently suffered from a deterioration in income, working conditions, and, in some cases, prestige (Altbach, 2002). Realities for academics in developing countries are, in general, significantly less favorable than for their colleagues in the North.

Institutional Environment

The working environment for most Third World academics is far different than what is the norm in the industrialized nations. While this chapter is not intended as an analysis of the infrastructures of academic institutions, it is necessary to point out that conditions vary considerably across and within countries. For example, India has a few academic institutions—such as the Indian Institutes of Technology, several management schools, and the Bhabha Atomic Research Centre— with facilities comparable to the average institution in the North, although not the very best. But the vast majority of Indian universities, colleges, and other academic institutions fall far below the level of the average postsecondary institutions in the North. While accurate figures do not exist, it is probably the case that 95 percent of Indian academics work in an environment that is well below international levels. The situation is only modestly better in China, which has a growing number of academic institutions that seek to compete on a global level in terms of research and teaching, but where the large majority of academics work in substandard conditions. In many developing countries, especially smaller nations, no academic institutions exist that even approach international standards in terms of facilities. Even large countries, such as Ethiopia or Nigeria, have few, if any, academic institutions that can offer the working conditions that would permit scholars and scientists to function competitively on an international basis. Even in fairly well-developed academic systems in relatively affluent countries such as Argentina, the physical facilities available to most academics are quite limited. What is surprising, and quite impressive, in the developing world is the ability of many academics to work effectively under such difficult circumstances.

The academic environment is characterized by inadequacies at all levels. The cost of maintaining up-to-date facilities has increased with the escalating prices of journals and books, and the complexity and sophistication of scientific equipment. In the 21st century, it is increasingly costly to stay competitive in world science. Further, all of these scientific products would have to be imported at unfavorable exchange rates and in an environment of financial scarcity.

As noted earlier, access to the Internet, while expanding in the developing world, remains inadequate. The infrastructure is antiquated and poorly maintained. Access to some of the major databases is limited by the high cost of accessing them. Few academics have work stations for themselves—computer use is rationed at many institutions.

In fact, many academics lack even a desk on which to place a computer, even if one were available. Office space is in short supply, limiting the possibility for academic work and for consultation with students and colleagues. Many academics have nothing but the books they use as texts or perhaps a few related publications. Without question, the physical infrastructure available to most academics is inadequate for scientific research and scholarship and barely adequate for teaching. Indeed, in much of the developing world, facilities are actually deteriorating due to financial shortages and the pressures of ever-increasing numbers of students.

Bureaucracy and Politics

Universities everywhere are bureaucratic institutions. In the North, the concept of shared governance is the norm, with the professoriate sharing or (decreasingly) controlling the key governing structure of universities. Professorial power has weakened everywhere, as academic institutions become larger and demands for accountability mount. However, academics' control over key aspects of the curriculum, the hiring of new faculty members, issues of instruction and evaluation, and related issues remains largely intact.

The same cannot be said for many universities in developing countries. First of all, the tradition of professorial power and shared governance is weak. In countries formerly under colonial rule, universities were founded with strong bureaucratic structures and firm controls to ensure loyalty and adherence to the norms of the colonial authorities. In other countries, academic institutions, which were often directly established by government, also lacked the traditions of faculty power.

In recent years, governments have been concerned about institutional stability, student political activism and unrest, and the risk that universities could become sources of dissent in society. These factors led to the buildup of strong bureaucratic controls and a lack of professorial power and autonomy. Even in Latin America, with its long tradition of formal autonomy for the universities, the academic profession has attained less control over their working conditions and over the structures within the institution.

Many universities in developing countries have become politicized, which has directly affected the academic profession. In developing countries, universities are important political institutions—not only do they train elites but they also play a direct political role as a forum for student political activism, dissident perspectives, and even mobilization of opposition activities. Especially in societies with unstable governments, universities often serve an oppositional political function.

In developing countries, two kinds of politics affect higher education: academic politics within the university and societal politics. Academic politics can be found everywhere—in departments, among colleagues, and in the university at large. In the North, while factions may be present in departments, institutions and units within them are generally not disrupted by politics or governed by political considerations. Seldom does the partisan politics in society at large intrude into the on-campus operation of the institution. In developing countries, politics is more prevalent at universities and is not infrequently a motivating force in academic policy decisions, the hiring or promotion of academic staff, and in other ways.

A number of factors explain the intrusion of politics into academe. In the developing country context, the university is an institution with considerable resources. In such a resource-scarce environment, the decisions made on campus, including the hiring of staff (faculty and administrators), student admissions, the creation of new programs, and so on, have broader implications. Universities in developing countries have a tradition of being politicized, having been involved in independence movements or other struggles. Politics has continued to be an element of campus life. The norms that in the North keep partisan politics away from the university are often missing.

In Latin America, for example, party politics sometimes permeates the election of academics to administrative posts. Candidates for rector or dean may stand for election backed by a political party or campus faction. Political influence is often felt in the appointment of professors and other staff. Occasionally, student admissions or examination results may be influenced by political considerations.

Universities everywhere are complex bureaucratic institutions. In developing countries, bureaucratic control, government involvement in academic decisions, and the politicization of all elements of higher education have been detrimental to the academic profession and to the preeminence of academic norms and values in higher education.

Academic Freedom

Not surprisingly, given the realities discussed here, academic freedom is often not well protected in developing countries (Altbach, 2001, pp. 205–219). The institutional protections common in the North are often missing—such as tenure or civil service status and institutional protections of academic freedom. A number of factors have combined to put professors in developing countries in a more vulnerable position than that of their counterparts in the North. The history of higher education in developing countries, as noted, is one of governmental oversight and bureaucratic control. Colonial regimes as well as postindependence governments worried about the political loyalty of the professoriate and of the university. There is, to a certain extent, a tradition of subservience in the academic profession of developing countries (Gilbert, 1972, pp. 384–411). Academic freedom is, in a sense, more than "academic" in many developing countries, because the writings and sometimes the teachings of professors may have direct political consequences beyond the university. The campus environment is often highly volatile, and professors may contribute to dissent on campus and in society. Protecting professorial freedom of expression and academic work does not receive a high priority from governments.

These limitations on academic freedom are a detriment to the professoriate. They create uncertainties concerning expression and research. And when professors step over an often undefined line, they can suffer serious consequences, from mild sanctions to loss of their positions, or imprisonment. In some countries, research, especially in the social sciences, is restricted. Publications are closely monitored, and professors who express views in opposition to government policy face problems. For most academics, the situation regarding academic freedom is not perceived as problematic. In the sciences few restrictions are imposed. Most academics are in any case involved exclusively in teaching, and classroom expression is seldom monitored. However, the lack of a respected culture of academic freedom has an impact on the intellectual atmosphere of the university.

Working Conditions

It is clear that, in general, the professoriate in developing countries works under much less favorable conditions than what is standard in the North. Again, there are significant variations—with a small proportion of academics at the top universities enjoying conditions similar to the North. Few classrooms have anything more than the most rudimentary teaching aids. Class size tends to be large, and in any case the almost universally accepted method of instruction is the lecture, with little opportunity for either discussion or questions. In some countries, the lack of laboratories or scientific supplies and equipment has deprived students of an essential component of scientific training. Rote learning has become the norm in many places.

Teaching loads, even for senior professors, are high by international standards, and academic staff spend more time in the classroom than do their peers in the North, although there are exceptions to this generalization—such as China. The practice of assigning advanced graduate students to assist professors is virtually unknown. Academic staff may spend 20 or more hours per week in direct teaching. Little time remains for research, course preparation, advisement, or other academic activities. Academic staff often have little control over what courses are taught. Differences do exist by country, rank, and institution, with academics at the most prestigious universities teaching less than their colleagues further down on the academic hierarchy. Junior staff often teach more than senior academics.

In a growing number of countries, academics are expected to engage in remunerative activities unrelated to their basic teaching. Consulting; extra instruction related to revenue-producing, noncredit courses, or other programs; extra-mural service; and other work are all increasingly part of the academic workload. These activities produce additional income for universities as well as for the individual faculty member. The traditional job of the professor is expanding to include entirely new kinds of responsibilities.

Remuneration

Undeniably, remuneration is a central factor in the life of academics. Without adequate salaries, professionals would be hard pressed to do their best-quality work. The gulf between the industrialized nations and the developing countries with regard to salaries is immense. Academics everywhere earn less than people with similar qualifications in the rest of the labor force. People do not become professors to get rich. Nonetheless, in most industrialized countries, it is possible for academics to achieve a modest middle-class standard of living on their salaries. There are variations by country, discipline, and rank, but generally, academic salaries are sufficient to live on in the North. In developing countries, however, with rare exceptions, this is not the case.

In many developing countries, a full-time academic salary cannot support what is considered to be a middle-class standard of living. This is almost universally the case for junior academics, but is also true of senior professors in many countries. Thus, in many countries, academics must hold more than one job. Their main appointments provide a portion of their income, but they must earn additional income from teaching at other universities, consulting, or even holding jobs in business or in service occupations unrelated to their intellectual work. In many countries, academics provide tutoring or other ancillary teaching in order to boost their income, even when such activities are proscribed by the university. In the industrialized world, professors also take on outside consulting in order to earn extra income. The difference in developing countries is that, without this additional income, academics could not survive. For many, there must also be other earners in the family.

Salaries do, of course, vary significantly across and within institutions. Private universities often pay higher salaries than public institutions— the majority of academics in developing countries work at public universities. Income is linked to rank, but in some countries professors engaged in research and graduate teaching can earn higher salaries. In a few places, there is additional payment for publications and other evidence of academic productivity. Salaries can be higher at the most prestigious institutions, in business schools, and other specialized schools. In some countries, academic salaries are not paid regularly, placing great strains on the affected academics, civil servants, and other public officials. The numerous part-time professors earn much less, in some cases just a token payment.

As a general rule, the prospects for improvement in the low academic salaries in developing countries are extremely poor. The implications of salary structure are significant, as the poor salary levels have led to brain drain. The best scholars and scientists in developing countries can earn many times their local salaries by relocating in the North, and many take this option. Few are able to devote their full attention to their academic work because of the need to supplement their incomes. Thus, an academic career in the Third World is less than a full-time occupation, even for academics who hold regular full-time positions. This has negative consequences for research and academic productivity, generally. When combined with the structural impediments discussed earlier, it is hardly surprising that the research productivity of academics in developing countries is so low. Salary structures also negatively affect morale.

WHAT IS TO BE DONE?

This overview of the academic profession in developing countries has provided a generally gloomy perspective. Although the outlook for improvement is not good, it is useful to point out some specific changes that would contribute significantly to morale, productivity, and perhaps most importantly, the quality of universities and other academic institutions. These suggestions are not complex—in some instances stating the obvious—but implementation will be a challenge in many countries.

- Adequate salaries and guarantees of a stable career path should be provided to at least a key segment of the professoriate who hold full-time positions at the main universities.
- At least at the top academic institutions, university facilities need to be upgraded sufficiently so that the most-well-qualified professors are able to pursue research and offer excellence in teaching.
- Procedures for involving the professoriate, along with administrators (and in some cases students), in academic decision making are essential to ensure that the academic staff have a significant role in the governance of the institution.
- The academic profession needs to be depoliticized—this would include the links between political parties and academics, the close ties between the professoriate and student activists, and the partisan nature of academic decision making and elections.
- Academic freedom must finally become a recognized part of university life, with guarantees protecting freedoms with regard to research and publications, teaching, and reasonable expression in the public sphere.
- The academic profession itself must develop a sense of responsibility with regard to expression and publication, especially on controversial topics.
- The academic profession must receive adequate training—the doctorate, for those involved in research as well as teaching; the master's degree, for those who are exclusively teachers; and for all, some exposure to training in pedagogical methods.

- Academics must be provided with the means to keep up with current trends in their fields.
- Great care needs to be taken to ensure that part-time and temporary academic staff are well qualified and provided with appropriate benefits.

CONCLUSION

This chapter has presented an almost unremittingly pessimistic picture of the current state of the academic profession in developing countries. Yet, what is surprising is that there are so many people working in higher education institutions who freely choose the academic life and who persevere under difficult circumstances. Fortunately, academic work in developing countries does have many rewarding aspects. Scholars are generally held in very high regard, and a professorship, even if poorly paid, is an occupation that has very high status. Learning is respected and those who possess knowledge are held in high esteem. Despite the circumstances described here, university life holds considerable attraction. It is, after all, the life of the mind, and those who are inspired to heed the call to intellectual pursuits will put up with many hardships to pursue an academic career.

Yet, as is clear from this analysis, the profession is truly in crisis. The consequences of continued deterioration in the conditions of the professoriate include not only neglect of one of the most highly educated and potentially productive segments of the population, but also the deterioration of higher education as well, since academic institutions cannot perform well without a committed, well-trained, and stable academic profession. In the context of globalization, developing countries require access to the wider world of science and technology, and the academic profession represents a central link to the international knowledge network. As the primary educator of future generations, the academic profession is in many ways the linchpin of development.

REFERENCES

Altbach, Philip G. (1998). Gigantic peripheries: India and China in the world knowledge system. In P. G. Altbach, *Comparative higher education: Knowledge, the university and development* (pp. 133–146). Greenwich, CT: Ablex.

Altbach, Philip G. (2001). Academic freedom: International realities and challenges. *Higher Education*, 41, 205–219.

Altbach, Philip G. (2002). How are faculty faring in other countries? In Richard P. Chait (Ed.), *The questions of tenure* (pp. 160–181). Cambridge: Harvard University Press.

Boyer, Ernest L., Altbach, Philip G., and Whitelaw, Mary Jean. (1994). *The academic profession: An international perspective.* Princeton, NJ: Carnegie Foundation for the Advancement of Teaching.

Castells, Manuel. (2000). *The rise of the network society.* Oxford: Blackwell.

Choi, Hyaeweol. (1995). *An international scientific community: Asian scholars in the United States.* Westport, CT: Praeger.

Clark, Burton R. (1987). *The academic life: Small worlds, different worlds.* Princeton, NJ: Carnegie Foundation for the Advancement of Teaching.

Evans, G. R. (2002). *Academics and the real world.* Buckingham, UK: Open University Press.

Gilbert, Irene. (1972). The Indian academic profession: The origins of a tradition of subordination. *Minerva,* 10, 384–411.

Scott, Peter (Ed.). (1998). *The globalization of higher education.* Buckingham, UK: Open University Press.

Shils, Edward. (1972). Metropolis and province in the intellectual community. In Edward Shils, *The intellectuals and the powers and other essays* (pp. 355–371). Chicago: University of Chicago Press.

Task Force on Higher Education and Society. (2000). *Higher education in developing countries: Peril and promise.* Washington, D.C.: World Bank.

Teferra, Damtew. (2002). *Scientific communication in African universities: External assistance and national needs.* Unpublished doctoral dissertation, Boston College.

Teferra, Damtew, and Altbach, Philip G. (2002). Trends and perspectives in African higher education. In D. Teferra and P. G. Altbach (Eds.), *African higher education: An international reference handbook.* Bloomington: Indiana University Press.

2

BIG CITY LOVE: THE ACADEMIC
WORKPLACE IN MEXICO

Manuel Gil-Antón

... And then the souls,
Not the delicious fruit, soft skin
Overflowing with sweet syrup
Tree-ripened in its own season,
But market fruit beaten into ripeness
By the force of brutal blows
—Jose Martí, "Amor de Ciudad Grande"[1]

Why do we prematurely force fruit into an artificial ripeness? Because in the big city, the hectic pace of a long overdue modernization means there's no time to wait for nature to take its own course.

In this chapter I shall try to show, first, how the formation of the academic profession in Mexico throughout the 20th century has almost always taken the course of trying to adopt elements derived from various models of university and academic life in other parts of the world, disregarding the fact that the country lacked the basic foundations to support the development of these models.

In the course of attempting to copy different models, valued independently of the background social conditions in their countries of origin, Mexico has created some original elements of academic life. While Mexican higher education finds itself in a situation filled with complex challenges and dilemmas, it can also claim achievements that form a heritage upon which to build a better future for academia as a whole and the academic profession in particular.

The second part of the chapter examines the impact that the major expansion of higher education, starting in 1960, had on the academic profession. The period of unregulated expansion was followed, in about 1990, by one characterized by differentiation. After a discussion of the challenges facing higher education in Mexico, the chapter concludes by proposing a series of actions to confront those challenges and help to recover the original traits that have evolved over time, as well as some alternative developmental approaches that may be more consistent with the future of the country.

A Century in Pursuit of Doctoral Degrees

For the Mexican academic profession, the 20th century began and ended with a call for more doctoral degrees. In September 1910, as part of the celebrations marking the first centennial of the war of independence from Spain, the government of Porfirio Díaz decided to create the National University (Universidad Nacional de México). A previously existing university, a legacy of the colonial period, had been closed for many years. In the interim, Mexican higher education functioned in several isolated professional schools (in medicine and law, among other fields) and at the Escuela nacional preparatoria (National preparatory school). The creation of a national university was the project of Justo Sierra, minister of public education at the end of the Porfirio Díaz regime. The idea was not new: It had been proposed, unsuccessfully, in 1881. The new university was actually the product of the merging of the professional schools, the Preparatoria, and other academic entities into a single institution, to which a projected École des hautes études would be attached.

The creators of the proposal faced a key problem: the teaching staff. The great majority of professors associated with the preexisting schools made their living from their professional practices and devoted little time to teaching. Without a full-time, highly qualified teaching staff, the university they had envisioned—modeled after the University of California, Berkeley—had very little chance of being realized (Garcia-Diego, 1996, p. 34). The people behind the new university could not count on having the political clout to upgrade the existing teaching staff by selecting those who were most committed to academia. They also lacked the economic and academic resources and the time to produce a full complement of full-time professors with doctoral degrees. The solution was to request from each school "a list with the names of up to one-fourth of the total number of professors, provided that they had put in several years of good service and considered their commitment as professors to be a key aspect of their lives" (Garcia-Diego, 1996, pp. 30–31). These professors were awarded doctoral degrees. In order to avoid authority problems, the ex officio doctoral degree was also granted to the directors of the schools. All these degrees were granted at the opening ceremony of the new university: "The purpose was to dignify and give more luster to the new institution, which would be launched with several nominal academic professors who were not even full-time professors" (Garcia-Diego, 1996, p. 31).

Almost 90 years later, in 1997, the vice-minister of higher education and scientific research in the Ministry of Public Education, referring to a draft of the Program for Teaching Staff Improvement (PROMEP), stated: "In summary, the issue is very clear for us. What characterizes the best universities of the world—let's say, Berkeley, Princeton, Oxford or Harvard? They have full-time professors with doctoral degrees. Thus, the way to enhance university quality involves a strong effort toward the formation of high-level professors, who must also be provided with the environment in which to perform their task on a full-time basis" (Gil-Antón, 2000a, p. 25). The program document states that "The figure of the full-time professor-researcher. . . will reinforce the academic dynamics by building the *foundation* for higher education" (PROMEP, 1997, p. 9). In 1997, according to this official document, only 30 percent of academic staff in Mexico had full-time appointments, and only 5 percent had doctoral degrees.

Consequently, it has been declared urgently necessary to reverse this "undesirable" situation as quickly as possible. PROMEP has established a number of goals for 2006. The proportion of full-time positions will be increased to 70 percent. At least 22 percent of these will require doctoral degrees, the rest master's degrees or a teacher training specialization. In brief, this program—the most important of the initiatives undertaken by higher education authorities in the 1990s— was intended to increase by fourfold the number of full-time professors with doctoral degrees, as compared to 1995, and to double the number of full-time positions. Moreover, everyone was required to conform by 2006 to the desirable profile—that is, to ensure that the 39,000 full-time staff, who now have only undergraduate degrees, complete their studies for the master's degree or for a specialization. The short-term challenge is immense, for this program is seeking to transform the academic degrees and working conditions of academic staff already hired—in other words, a whole re-engineering process.

There is, of course, an enormous disparity between the situation in 1910 and the current context: the difference between handing out honorific degrees just to feign being a modern university and the current planning effort with true professional development programs for the teaching staff. Common to both cases, however, is the intention to modify the profile of academics in order to make them conform to the appropriate characteristics of what a university "should be," to as great an extent as possible.

In both cases, the national institutions, when compared to the model university, were found to be inadequate, therefore making modification an urgent need. This leads to a corrective approach based on designing strategies to emulate the model. The conditions are radically different, but the common approach is to approximate as closely as possible the formal aspects of the model—that is, Berkeley—without meditating on the specific circumstances beyond our borders that made its existence possible.

Put somewhat bluntly—and the nuances will be explored throughout the chapter—it sometimes seems as if we confuse the effects with the cause: if we agree on certain ideal characteristics, then the concrete processes needed to support them will automatically follow, along with the social rules that will serve as reference points for the actuating parties. Unfortunately, this strategy will not work, and with this paradox we began and ended the 20th century. The coincidence, in my judgment, is not trivial.

CHANGE AND PERMANENCE

The times in which we live, in transition between two centuries, seem to show that Heraclitus was right: *everything changes,* the essence of being is change, unceasing mutation, the true opposite of permanence—a pattern appreciated by Parmenides as the very nature of being. This sensation of an unrestrainable process of variation in all aspects of life, when nothing stays still and movement is the only permanent thing, is not foreign to Mexico.

It has been asserted that the 20th century, beyond chronologies, ended with the fall of the Berlin Wall and it was then that we actually commenced a new era worldwide. Perhaps. But in Mexico, future historians will probably recognize Sunday, July 2, 2000, as the date that marked not the beginning of change but its culmination. On that day, for the first time in more than 70 years, a presidential candidate from the opposition was elected into office.

In 1910, at the outbreak of the Mexican Revolution—two months after the opening ceremony of the National University—the basic demands were for *sufragio efectivo y no reelección* (effective suffrage and no reelection). At the federal level, 90 years had to pass before this principle was truly realized and the value of electoral democracy was insituted at the highest levels of the political system.

How can we understand the changes in the environment of the academic profession in Mexico, as well as future challenges, without considering worldwide and national trends? When observed from a historical perspective, more in keeping with what Braudel called the "long term," the apparent exclusivity in the ebb and flow of everything is dissolved, and we may then appreciate the continuity, persistence, and common themes that link change to permanence.

This chapter has a core idea: Underlying the great changes in Mexican higher education over time is the continuity of persistent attempts to emulate the characteristics of *some model*. In other words, throughout the 20th century the history of higher education in Mexico and the academic profession can be understood largely as efforts to replicate a model—indeed, various models. Successive attempts were made to copy external models in a hurry and without considering the social, economic, and cultural factors that made them possible in other places but that are not reproducible at will. As José Joaquín Brunner has taught us, efforts to adopt external modalities have generated a copy, each time. Nevertheless, each copy is an *original* (Brunner, 1987).

OF MODELS AND STARS

When Mexican higher education policymakers dictate the university model to follow, they tend to make generalizations based on the situation of certain universities in other parts of the world, as if the peculiarities of those institutions were typical of the higher education systems in their respective countries. For example, they state that doctoral degrees and full-time working conditions are the universal rule in the United States. This is not true—it certainly occurs, but only in a subgroup of universities. Based on 1993 data issued by the National Study of Postsecondary Faculty, Finkelstein, Seal, and Schuster show that, in general, there is no universal rule for the ratio of academic staff in the United States with Ph.D.s and tenure to those without, and that in the last few years, especially regarding the new generation of academics, tenure has become less frequent (Finkelstein, Seal, and Schuster, 1998).

The conditions of the "golden age" of academia in the United States seem less sustainable than before: For several years now, experts from that country have talked about the "Latin Americanization" of the academic profession when referring to the significant increase in the number of professors hired as part-timers at American universities.

Therefore, it is possible to postulate another paradox: In Mexico, it is believed that we should follow what seems valid for higher education institutions in developed countries. Consequently, the decision has been made—for example, in 1997—to take on the characteristics of American universities as soon as possible, at the very time the latter were actually showing a change in their classic academic profiles through diversification, the impact of fiscal and financial constraints, accountability processes, and increasing technological innovation.

These models, then, look like the stars. They are so distant that the light we now receive comes from earlier times, when they were alive and shining: Today the situation may be quite different.

It is appropriate, at this time, to make an important observation: The trend of adopting models generated in other parts of the world, as points of reference for countries coming late to the modernization process, has never been a simple phenomenon. It was never just a case of attempting to copy what was happening in other parts of the world, but rather something more complicated. In the closing decades of the 20th century, Mexico experienced pressures at various levels of society from the need to ensure access to the global parameters of modernization, which frequently involved obtaining credits or financial rescue schemes and meeting the conditions contained in international commercial agreements. These developments have affected public spending and the logic followed for the allocation of fiscal resources.

Perhaps because of Mexico's entry into various multinational organizations or its commitments to international funding agencies and involvement in exchange agreements and discussions about development, higher standards have been imposed on higher education— without enough thought being given to the country's ability to implement the changes. To some extent, this situation has given rise to confusion, falsification, and simulation of change. At the same time, however, the actions were taken in response to the needs and expectations of the parties involved. I refuse to believe that the flow is exclusively in one direction. Rather, to my understanding, it is about a complex convergence between internal and external factors, the results of which still need to be clarified.

In any case, the prevailing trends regarding higher education elsewhere in the world—necessary for benchmarking—need to be examined in the light of current data, original research, and recent regulatory efforts—so as to avoid generalizing from stereotypes that are no longer valid. It is necessary to have a wider comparative perspective and to take into account, for instance, the European and North American situations (Altbach, 2000; Enders, 2001), as well as the diversity of situations presented in this volume.

THE ACADEMIC MARKET: GENERAL FIGURES

In 1960, the first year for which enough information exists, the total number of academic positions in Mexico was close to 10,000—almost all of them (at least 95 percent) part time. Even in the absence of official data in this regard, it can be assumed that the majority of academic staff had educational backgrounds consisting of professional undergraduate studies.[2] This academic staff served approximately 78,000 students, or 2.7 percent of the higher education age cohort (20–24 years of age). They worked at 50, mainly public, institutions (Gil-Antón et al., 1994, p. 24; and Gil-Antón, 1996, p. 310).

Today, the situation presents a striking contrast: By 1999 higher education enrollments had reached a little over 1.8 million students at 1,250 higher education institutions (515 public and 735 private).[3] Even with the greater number of private institutions, enrollments at public institutions constitute 70 percent of the total number of students. The participation of the age group is close to 18 percent—a more than sixfold increase in 40 years—and the total number of academic positions is 192,000. Full-time positions have risen to 29 percent and half-time positions have risen to 9 percent; thus, the proportion of part-time staff has been reduced to 62 percent. With respect to academic staff assigned to graduate studies, at the national level only 35 percent have full-time contracts and 8 percent have half-time contracts. Part-time staff, even at the graduate level, still comprise more than half the total (57 percent), which is a function of the importance of professional specialization in graduate studies—for example, in medicine (ANUIES, 2000, pp. 88–94).

Eighty-two of every 100 academic positions are devoted to undergraduate studies, 9 to teacher training schools, and another 9 to graduate studies. Of staff holding full-time positions in undergraduate studies—approximately 50,000 professors—26 percent have a master's degree and only 6 percent have doctoral degrees. Therefore, we may conclude that the remaining 68 percent have undergraduate degrees or advanced professional specialization studies degrees (ANUIES, 2000, p. 90).

CREATING NEW ACADEMIC POSITIONS

Going from 10,000 academic positions in 1960 to 192,000 at the end of the 1990s meant a significant expansion of higher education opportunities in the second half of the 20th century. The expansion averages out to almost 13 new positions added every 24 hours. However, this growth in turn reflected other profound social changes—industrialization, urbanization, and the growing intricateness of social life.

The growth in the number of academic positions in Mexican higher education and the social effort it involved are better understood if they are related to some contextual data. The context was that of a country embarking on an overdue modernization and a society marked by great inequalities and deficits in basic services. In 1940, according to census data, 90 percent of Mexican adults over the age of 15 corresponded to the category of "extreme educational poverty": 58 percent were illiterate; 25 percent had some reading, writing, and elementary calculation skills, not learned at school; and the remaining 7 percent had not completed the mandatory six years of elementary school established at that time by the Constitution. In 1980, for the first time, Mexico reached a figure slightly below half: 48 percent of adults in the country continued in the category of extreme educational poverty (Boltvinik, 1995).

By 2000, education authorities estimated that 35 million adults— more than one-third of the total population—lacked a basic education (defined in 1993 as nine years of schooling): 18 million lacked *secundaria* (three years after elementary school); 11.2 million lacked *primaria*, or elementary school; and 6.6 million were illiterate (*Bases para el programa*, 2000, p. 67). In 1970, only 1.09 percent of adults had completed undergraduate studies or higher levels of education; in 1980 this group comprised 2.7 percent and in the early 1990s, 2.9 percent (Boltvinik, 1995).

The decision to increase the number of higher education positions involved the accelerated recruitment of academic staff, which led to a number of very important developments: the expansion of educational opportunities, the extension of social mobility to new sectors of the population, and the opening of prestigious job opportunities for a large number of "pioneers" who had just arrived at the universities seeking professional development in various fields. The following imaginary dialogue would not have been unusual during the 1970s or 1980s: "Wouldn't you like to be a professor?" "But I am just finishing my undergraduate law degree." "It doesn't matter—we have so many new students and not enough professors. You are about to finish your studies and could surely handle some beginning courses." "Well, the thought had never occurred to me . . . but what would I have to do?" "Come in early tomorrow, and we will give you the course syllabus, your class schedule and room assignments, chalk, and an eraser. Good luck, colleague!"

The span of 40 years to which I referred saw a number of significant developments. The pace of creating new academic positions was uneven during the period. The predominant model of the university underwent changes that undermined its viability; the country transformed its development policies, passing in a very short period from a closed to an open economy responsive to international factors in a way that would have been unthinkable before. And all this occurred at a time of significant worldwide changes. This series of factors has had an impact on the working conditions within the academic community. Without a clear understanding of these developments, the present reality of the Mexican academic profession, as well as its future challenges, would remain obscure. Before addressing the problems by once again proposing a new model to copy, a historic perspective is essential.

THE FORMATIVE PERIODS

Broadly speaking, we can divide the evolution of the academic workplace in Mexico into two periods. Between 1960 and 1990 the expansion of higher education proceeded on the basis of a model focused on the reproduction of professionals, and in response to increasing social demands motivated by expectations of social mobility. By the end of this period, parts of the system went into crisis as a result of the failure of strategies based on this model. It appeared that the expansion was carried out without adequate academic oversight, given the policy's effectiveness in terms of the political legitimacy and stability of the system as a whole. The process was, in general, a reactive, not planned, growth drive (Metzger, 1987).

While pointing out these problems, we should not overlook the fact that at the margins during these times, alternative forms of higher education were also developed focusing on the disciplines and the effort to reach higher levels of scientific and technological activity. This modernization of academic activity is called the "creation of professionalization enclaves" by Brunner in his studies on the Latin American university (Brunner, 1987).

After 1985, with the economy in crisis, there was a major change in the conduct of Mexican higher education. However, since the 1990s, a clear change of direction left nonregulated expansion behind and focused on differentiated contexts, evaluation, and monitoring with the goal of reorganizing the system.

The First Period: Unregulated Expansion (1960–1989)

Between 1960 and 1970, at the beginning of the period of Mexican higher education expansion, an average of 4.12 new positions were produced each day. From 1970 to 1985, when the pace of growth had quickened, on average 12.92 new positions a day were created. The second half of the 1980s, on the other hand, showed a remarkable contraction down to 5.08 new positions a day (Gil-Antón, 1999, p. 35).[4] This period saw the extension and exhaustion of policies based on the higher education model oriented toward the reproduction of professionals, whose central actor was the *catedrático*.

Before the 1960s, in an indication of the limited scope of higher education in Mexico, it can be said that academic activity focused on teaching courses. *Catedráticos* were leading professionals in their fields—as was true at the beginning of the century—who devoted some time to teaching, an activity valued more as a distinctive status symbol than merely as a source of income. "I am a *catedrático* at the university"—whether at the National Autonomous University of Mexico or the corresponding state university—was a statement of pride and status. At the beginning of the expansion in enrollments, obtaining new professors by recruiting outstanding students and gradually initiating them into the duties of teaching assistants began to be inadequate: The output could not meet the demand for new professors. And when the rate of growth reached almost 13 new positions each day, recruiting methods had to change.

In 1992 a pioneering study of the academic profession in Mexico that surveyed by questionnaire 3,764 professors was carried out (Gil-Antón et al., 1994). The study yielded a profile of these academics when they got their first teaching opportunity—that is, when they formally assumed responsibility for one or more undergraduate courses. Their average age was 28 years, but a significant number (28 percent) were hired before the age of 24—that is, within the typical age range of an undergraduate student. Seventy-four percent began their academic careers before the age of 31. Only 9 percent came from backgrounds in which both parents had university degrees. Seventy percent were the first in their families to attend university and were also the first to achieve a prestigious occupation—that of a university professor. Thirty-five percent began teaching before having obtained an undergraduate degree. Forty-nine percent had only an undergraduate degree at that time. Only 12 percent began their academic careers with a degree higher than the one toward which their students were working. Eighty out of every 100 were hired at the same institution at which they had studied. In 63 percent of all cases, the respondents had no previous teaching experience of any kind, and 93 percent did not have prior experience in research. Sixty-eight percent combined teaching with another job, frequently related to their own professional field, since the vast majority (75 percent) had part-time academic contracts.

These data indicate the departure from the recruitment pattern that existed previous to the expansion. Of the *catedrático* model, conceived as that of a distinguished professional with some teaching responsibilities, very little is left. The contrast is enormous between Mexico and the developed countries, where the doctoral degree is essential in order to begin one's academic duties and yet not sufficient to ensure a successful career, as additional qualification procedures in addition to the degree are required. Was the situation in Mexico caused by errors, carelessness, or the aberrations of underdevelopment? No—rather, from a sound sociological perspective; we should, as Weber said, pose the right question with regard to the phenomenon: Why was it so and not some other way around?

The political decision to expand the number of higher education places to accommodate the growing number of students graduating from secondary schools was a result of the social modernization that had occurred in Mexico in the previous decades. Today, the average education level in Mexico is 7.7 years. In the period of accelerated expansion, the average was probably less than 4 years. Thus, the gap between the average person's education level and that of an undergraduate student was, at that time, more than 12 years. Recruiting those young undergraduates into the academic profession was the only option at the time. As pioneers, they undertook the mission of filling the newly created positions in university classrooms and laboratories.

In comparison to academics in some other countries, the pioneers did have training deficiencies: A weak background in their respective disciplines, scant professional grounding in their respective fields, and slight qualifications for teaching and almost none in research. Nevertheless, they were taking part in the building of a higher education system that has contributed significantly to the transformation of the country. Moreover, they were helping to break down the entrenched hierarchy and didactic rigidity of the older generation of professors. The pioneers were forced to innovate out of necessity, learning to teach in crowded classrooms, building up the institutions while creating their own working conditions. The strategy may have been to copy a model, but some of the results were original and not all were negative. Furthermore, the professionals and scientists of that pioneer group are now part of the patrimony that has led the country to new levels, meeting challenges along the way. The Mexican academic profession has, to be sure, experienced areas of confusion and instances of inconsistency, but there have also been real achievements that will lead to the emergence of new traditions.

One aspect of the recruitment of these professors, noteworthy since it is not tied to the developmental structural conditions of the country, has to do with its initiation period. When interviewed in 1992 and asked to recall their first year as professors, the respondents were asked whether they were assisted by another, more experienced, professor (i.e., worked as an assistant to that professor) or whether they took up their jobs alone, with full responsibility for their actions. The results were very significant: 71 percent answered that they started with sole responsibility for their classes. This relatively lonely, isolated entry and initiation into academic life was experienced by 46 percent of professors who began their careers before 1960, 57 percent who began in the 1960s, 71 percent who began between 1970 and 1985, and 79 percent who began between 1986 and 1992. While other characteristics of individuals who became academics are attributable to the conditions of the country at that time, the manner in which so many were initiated was in response to the compelling logic of growth.

Of the approximately 140,000 professors currently teaching at universities and technological undergraduate institutions in Mexico, some 70 percent went through this process at the beginning of their careers (Gil-Antón, 2000b). While the exact figures are not available due to the lack of reliable official statistics—a real obstacle to research on the academic profession—probably about 100,000 professors still active today went through this kind of process. Most of them now have about 15 years of seniority on the job, with an average age close to 46 years and professional development and experience profiles quite different from the ones with which they began their careers. According to 1992 data, 65 percent of this group had educational qualifications beyond the undergraduate degree level: 16 percent of them had a graduate specialty, 37 percent were at the time studying for or had already obtained master's degrees, and 6 percent were on their way to completing Ph.D's. Doctoral degree holders accounted for approximately 6 percent at the time of the interviews. Almost two-thirds had tenure in their jobs, and 40 percent had achieved full-time positions (Gil-Antón et al., 1994, pp. 136 and 138).[5]

Over time and varying by institution, the knowledge areas, disciplines, and regional distribution of members of this pioneer group, who shared some general traits upon entry, have changed. Attempts have been made to profile the backgrounds of members of this group—broken down by career concentration in teaching, research, or university management (Villa-Lever, 1996). Several types of academic backgrounds have been identified for professors, researchers, and professionals involved in academia (Gil-Antón, 1997). In the first studies of this group of professors that were conducted in the early 1990s, attempts were made to divide the sample into two groups— those fully devoted to the university and those only marginally involved (Gil-Antón et al., 1994, p. 163). Similarly, in the Mexican portion of the comparative study sponsored by the Carnegie Foundation for the Advancement of Teaching, with a national sample of 1,027, a distinction was made between individuals fully and marginally involved in academic activities (Gil-Antón, 1996, p. 320). Origin may not be destiny, yet at the same time, this group is marked by its origins. Still, it continues to evolve.

During the first 20 years of this period of unregulated expansion of the academic workplace in Mexico, the salary and prestige associated with academic activity remained comparatively high. However, in 1982, when Mexico entered a period of severe economic crisis, the purchasing power of salaries dropped: it is estimated that by the end of the 1980s salaries had lost at least 60 percent of their purchasing power. This phenomenon was not a minor one, but the "solution" adopted was still more important. As salaries dropped and the financial resources of public institutions were exhausted, a strong relaxation of work ethics occurred in the context of uniform retributions: Since individuals who continued to meet their responsibilities with regard to courses and other tasks were paid the same as those who took additional jobs and performed their duties with only minimal diligence, professional commitments were eroded to such an extent that the following phrase was frequently heard in university corridors: "If they pretend to pay my salary, then I will pretend to work."

This process of structurally induced simulation of work by academics did not occur to the same degree at all institutions or in all disciplinary areas, but its impact was felt throughout the system and has changed the conduct of academic work since the 1990s.

Before concluding the description of the first period, a peculiarity of the Mexican system needs to be explained—namely the definitions of the terms *full time* and *part time*. The reader will need this awareness in order to understand some of the characteristics of the second period. This aspect of Mexican higher education constitutes a serious limitation on research on the country's academic profession.

Throughout the years a large number of types of academic appointments officially categorized as full time have been created. In other parts of the world a full-time position may unequivocally describe that of an academic whose sole job is in higher education, with a certain balance between teaching, research, and service. In Mexico, a full-time academic can fit this description, but the term may also describe professors who only teach—perhaps as much as 27 or 30, or even more, hours a week—devoting the rest of their "free hours" to advising students and preparing lectures. Others may teach nine hours a week in addition to their main jobs (i.e., their main source of income), which might be working as professionals outside academia. They receive full academic base salaries and other benefits— such as sabbaticals and medical insurance. The salary is certainly low, but an academic in this group has a *definitividad* (long-term contract), which provides, as it were, a safety net in case the situation with their main work is jeopardized.

It would be beyond the scope of this chapter to outline all the different meanings of the term *full time* in Mexico. However, a couple of other examples may illustrate the range. For many years, a full-time job at a state university covered only 20 hours a week. This was due to the meager salaries offered and intended as a means of allowing professors to have two full-time contracts—at the same institution— to improve income levels. At the other end of the spectrum, many professors today have several part-time contracts, in different institutions, adding up to many teaching hours. Humorously, these professors are known as "ultra-full-time professors" or "taxi-professors," due to the fact that they travel back and forth to teach classes in as many places as possible in order to earn enough money.

At what rates do these and other variations in academic work occur? It is impossible to know because official sources report the number of posts per institution, but do not give information on the number of people who hold these posts. This appalling accounting system has not been modified, despite repeated demands from researchers for over 10 years.

The Second Period: Differentiation (1990–2001)

Within the field of higher education in Mexico, the crisis of the 1980s refers to the abrupt decline—after 1982—in salaries and institutional funding. However, the crisis also referred to the general economic crisis, which could not be solved by simply waiting for better times but required a transformation of the country's basic development model.

The crisis in the early 1980s was not a simple problem of "cash flow" in the national finances, as some observers claimed. The country was not just undergoing a temporary adjustment in economic development and social expenditures. Also at work were factors as essential as the role of the state in the economy, the place of the market in national development, and the proper objectives of social expenditures based on tax revenues.

In concert with (or coerced by, in the view of some observers) global trends, the dominant approach for managing the country changed significantly. To cite a couple of examples: Before, Mexico's was basically a closed economy. The opening up of the economy witnessed at the end of the 1980s was actually a surprise, as was the reduction in the role and functions of the state. Mexico went from having a political system full of authoritarian "certainty" to having a competitive, open, much more equitable election process, culminating in the federal election of 2000 and the democratic "uncertainty" of the new government (Gil-Antón, 2000b).

In the general context of the mid-1980s the Sistema nacional de investigadores (National system of researchers, or SNI) was created for the purpose of retaining the loyalty of the members of the most established academic sector in Mexico—those at greatest risk of migrating to other markets, national and international, to obtain better living and working conditions. Based on periodic evaluations, the SNI awarded additional increments of income to certain academics. A whole set of attributes typifying a professor committed to research was established: Professors recognized by the SNI were, and still are, part of what are known worldwide as research universities. A new model for the academic profession began to emerge in Mexico.

Since the 1990s the logic of income differentiation, with periodic qualitative evaluations—or quantitative evaluations, as others charge—of academic work, has become firmly established throughout the system. To be sure, base salaries have lagged greatly behind inflation, and general raises to recover purchasing power have declined. However, other sources of income have opened up: residency stipends (*beca a la permanencia*), incentives for outstanding work, productivity bonuses, and extra payments for graduate degrees achieved. These financial awards are not open to all staff but usually to a certain proportion—close to 30 percent of full-timers at each institution.

As a crisis plan to address the serious problems of declining salaries and lack of attention to varying aspects of academic work, this strategy mobilized an important sector of professors looking for recognition and adequate income levels. For the educational authorities, besides resolving or alleviating the problem of academic salaries, this policy created incentives for professors to adopt the universally validated professional model. That model was very research oriented and assumed a high degree of formative training and participation in international scholarly exchange and publication networks. Moreover, this program would be incorporated into new positions, from the start, which would provide guidance and direction. With regard to the production of new positions, the rate of growth has not subsided—it is again close to 13 new positions a day.

Having remained in place for more than ten years, the strategy can no longer accurately be described as a crisis plan. Moreover, as the plan has become well established, its "perverse" effects tend to nullify its positive effects, of reactivating academic activity.

Income differentiation and the relevance of research productivity are discussed all over the world, but in the Mexican case the debate has taken on greater intensity. Some examples will be illustrative of the Mexican context regarding income differentiation. Today, some professors near the top of the income scale earn a base salary of $1,300. On top of their base, they may receive $550 a month for having a Ph.D., another $500 from their residency stipend (which prevents them from working outside the university for more than eight hours a week), another $550 for teaching performance, perhaps another $300 for outstanding academic background—for a total of $1,900 a month outside of their contracts. The strategy is most well established at the Universidad Autónoma Metropolitana (UAM). Furthermore, if professors are at level I in the SNI, they receive another $600 a month, for a total non-contractual income of $2,500. In other words, of a total monthly income of $3,800, 66 percent comes from additional funds that do not affect fringe benefits—especially the pension. In retirement, a professor in Mexico receives only a fraction of his or her base salary and nothing else.

This hypothetical case presents one view of this new regulatory approach in academic life, but at this point it may be useful to offer a wider overview. The public institutions include, on the one hand, the federal institutions, which receive their budgets after negotiation directly from the federal government—the Universidad Nacional Autónoma de México (UNAM) and the UAM. On the other hand, there are the state universities, which are linked to the Vice-Ministry of Higher Education and Scientific Research within the Ministry of Public Education.

In the first case, that of federal institutions, the amount of money allocated for income differentiation is larger and the program is open to all full-time academics who satisfy the conditions established by the evaluation process. The case of UAM has already been described. As for UNAM, it offers four levels of additional income to its full-time academic staff. These range from 45 percent of base salary, for level *a;* 65 percent, for level *b;* 85 percent, for level *c;* and 105 percent, for level *d.* Once again, if a full-time academic attains level *d* (the highest) in an evaluation, which takes place every three years, he or she obtains more than 50 percent of total income in additional payment from his or her own institution. If this academic belongs to level III in the SNI (a very high level), he or she obtains an additional approximately $1,400—in other words, more than two-thirds of the individual's income comes from payments over and above the base salary.

In the case of the state universities, reward programs generally allocate the equivalent of 1.5 additional minimum salaries to 30 percent of full-time academic staff. This amount may be distributed uniformly to 30 percent of the full-time staff, or it may be distributed differentially, according to rules decided by the institution itself, which can even choose to allocate some of its own resources. In the Universidad Autónoma de Baja California, for example, there are five levels of supplemental income: At the lowest level of the top rank, a faculty member receives 85 percent of his or her total income through their contractual salary. However, at the highest level of the top rank, the additional money represents 34 percent of the total income. Additionally, if this professor belongs to level I of the SNI (the lowest level) he or she earns an additional $600, which means that half of the total income would come from sources outside the basic salary.

Given the wide variations and insufficient data, it is impossible to give a more complete picture of this phenomenon. However, there appears to be a clear trend: the "regular" salary—the portion subject to collective hiring practices—tends to decrease as one moves forward in an academic career. In the end, this implies that the academic profession in Mexico has a significant degree of income instability. At issue is not only maintaining the productivity required to attain internal and external levels of supplementary salary. Another issue is the fact that the financial conditions of institutions are subject to change, and a time may come when the sums allocated to different levels will need to be reduced in order to secure the financial viability of the institution.

Within the private sector it is necessary to distinguish between two classes of institutions. Elite institutions, such as the Instituto tecnológico y de estudios superiores de Monterrey or the Instituto tecnológico autónomo de México, offer very attractive salaries to their full-time academic staff and certain additional rewards that do not exceed 25 percent of their salaries. On the other hand, the rest of the private institutions basically pay by the hour for teaching and offer minimal benefits.

Since these arrangements have fairly recently been introduced into the academic income system, there is great diversity. However, a clear trend seems to be emerging in the public sector: a sharp income differentiation, based on periodic evaluation, as faculty move forward in their academic careers. In the private sector, and within the subgroup of elite institutions, the trend appears to be to offer a good salary, differentiated by discipline, but with a low ratio of full-time posts and hardly any tenure in the formal sense of the term.

Income differentiation, continuous evaluation of all academic products and performance, concentration on areas of work that result in rewards and abandonment of the ones that, although important, do not result in bonuses are the distinctive features of the second period. And in the judgment of experts in these matters, these arrangements have had an unforeseen effect: As the emphasis has been placed on evaluation of research and its products, along with the rewards to high levels of academic formation, the focus on teaching has been weakened—especially at the undergraduate studies level—although it is without a doubt the core activity of higher education. "You only teach, you don't do any research, you don't have a doctorate? You're not doing well at all, colleague."

Before the current drive toward speedy completion of the doctorate, for a number of decades there had already been a trend under way among academics toward the achievement of higher degrees, with variations by disciplinary ethos: It was more common to find persons with doctorates in physics, mathematics, or biology (almost 30 percent in each field in the 1992 Carnegie sample), than in the areas of accounting, management, law, medicine, or veterinary science (almost none in 1992). The latter earned status by filling very prestigious positions or specialized positions, which evolved in a very different direction from traditional graduate studies (Gil Antón, 2000a).

Today, having a doctorate and being a member of the SNI have become prerequisites for obtaining research funding. The doctorate has become a quality indicator denoting excellence in academic background. In fact, at UAM, a monthly bonus is granted just for having completed the degree (regardless of what is done with it). To illustrate the trend, in areas such as veterinary medicine, for example, a doctorate in "meat sciences" has been approved. Invention knows no limits, apparently, when it comes to imitating—in the interest of survival—the formal aspects of a presumably universal model.

Professors who began their careers before the introduction of differentiation into the public sector have lived through three different phases in their own working history:

> [The first, phase was one] of relative salary abundance and stable conditions, combined with low standards of academic regulation as regards access; then, they experienced an abrupt drop in salary levels, with a sharp reduction in the already scarce internal regulation; and finally, they faced the need for reorienting their activities and profiles, in many cases, in order to adjust to conditions of income differentiation. (Gil-Antón, 2000b, p. 164)

Looking at the academics who entered the profession during the economic crisis of the 1980s, a number of differences can be discerned in comparison with their colleagues from the earlier periods. As opposed to the earlier generation, this group had to compete strongly to obtain their initial appointments, as positions were scarce and the number of people interested in filling them had increased due to general employment constraints during that period and subsequent ones. As for professors whose careers started in the 1990s, the qualifications required have definitely increased, and expectations of stable employment are not what they once were. These professors arrive at institutions that impose, both at the outset and later, increasingly competitive conditions. It would be interesting to examine relations between the already established academics who have access to additional income and the new generation of young professors who are now joining academic institutions.

These considerations involve mainly public-sector higher education, which is still the major academic workplace in the country. In 1970, academic positions at private institutions accounted for 15 percent of the national total; today they account for 30 percent, 80 percent of them part time (ANUIES, 1970 and 1999). Nevertheless, the relevance of the Mexican private sector grows every day, as middle-class young people unable to gain admission to the prestigious public institutions or to other public schools have been joining the sector of society that has traditionally attended the elite private institutions (Levy, 1995). In previous decades, eight out of ten new academic positions were generated in the public sector. However, during the 1990s six out of ten were created in the private sector. Not surprisingly, these were mainly part-time positions because private institutions lack the capacity to support a greater number of full-time positions. These institutions are very market dependent and, except for a few elite institutions, devote very little attention to research activities.

A number of questions remain unanswered due to the lack of well-supported research. How do conditions vary if we take regional, institutional, and disciplinary differences into account? How will the retirement of large numbers of senior staff be managed in conjunction with the recruitment of those who will take the retirees' places in the coming years, in view of a very probable increase in higher education demand?

Today, we know much more about Mexican academics than we did during the 1980s. More than a dozen doctoral theses were produced in the 1990s on this subject—along with many books, articles, and studies. In 1999, the Sociedad para el estudio de la profesión académica en Mexico (Society for the study of academic profession in Mexico) was created, made up of about 30 researchers. The first steps are under way to formulate the state-of-the-art instruments that will allow us to recognize developments in Mexican higher education. However, the country still lacks the ability to research and answer the above questions pertaining to the academic profession.

Higher education is now entering a period characterized by very different rationales for the regulation of academic work. The earlier period, distinguished by a corporatist and bureaucratic logic, has given way suddenly to one ruled by an individualized orientation toward obtaining the greatest benefits within the institutional wage scales (Ibarra, 1998 and 1999). In anthropological terms, we have gone from conceptualizing university work as part of the construction of democracy in the country—investing in institutional construction with a collective *ethos*—to the concept that each individual needs to comply with the requirements in order to "be somebody." This need comes from the feeling of being classified as a "nobody" for not having a doctoral degree or publications in international journals and books, or for having a particular interest in teaching undergraduate courses (García Salord, 1998 and 1999). From another theoretical perspective, this transition is seen as the restructuring of the Mexican university into an institution that distinguishes between who should be "included" and who should be "excluded" on the basis of a set of indicators derived from the classical *ethos* of the hard sciences and the specific traits of an individual's cultural capital (Casillas, 2001).

Dilemmas, Challenges, and Possibilities

In the context of the presidential election of 2000, ANUIES—the Asociación nacional de universidades y instituciones de educación superior (National association of universities and higher education institutions)—outlined a number of strategic lines of development for higher education in the 21st century. What can be expected for the near future? The figures are sobering.

By the year 2006, if things go right—that is, if the most positive scenario occurs—enrollments will have passed from their current total of nearly 1.8 million students to 2.8 million: that is, one million additional students in just six years. Participation of the relevant age group would rise to 26 percent from its current figure of 17.7 percent.

As for professors, by the year 2006 there will be 291,000 positions—that is, almost 100,000 more than today. These projections are based on the current student/professor ratio (9.5 to 1) and current distribution of contract types. This would mean breaking all records on average daily production of new positions: By 2006, the figure would reach almost 35. With an improved student/professor ratio (12.5 to 1) and a reduction in the number of professors hired by the hour from the current 62 percent to 55 percent, 30,000 new positions would be necessary (ANUIES, 2000, pp. 125–136). What will actually occur? Only time will tell.

For demographic reasons, the age group that will grow the most in the next 25 years is the one associated with secondary and higher education. We are, therefore, at a point in time when whatever is done about education will impact a majority of the population. Therefore, it is important that the academic profession address the core challenges and dilemmas of our time.

During the last four decades of the 20th century, the academic workplace in Mexico has undergone several transformations. We have seen how, through the years, the strategy has been to try to force the "ripening" of Mexican academics in a process characterized by speed, hasty recruitment, abrupt changes of direction, and an endless succession of models. Perhaps what has really marked these times is the double task of forming a national academic profession while building the institutions and developing the new professionals required by the country.

If learning from history means not repeating it, it is necessary to re-examine the models—the solutions and development alternatives—that, in an increasingly open world, will continue to exert their power of attraction. Throughout this chapter, the goal has been to show that public policy related to the Mexican academic profession has emphasized not the guidelines and formation processes, but rather the specific features of the models being used. It does not seem appropriate, or possible, to isolate ourselves from the world and its influence, especially in these times of the elimination of frontiers, but it is necessary to take charge of what we have learned.

To my understanding, the first task is to recognize the need for a clear-eyed view of the past, which will enable us to understand what has been achieved, salvage the innovative results born of Mexican attempts to emulate different models, and examine the specific aspects of the Mexican academic profession, in the context of international trends. For this purpose, it is necessary to take advantage of political changes and establish a popular consensus on the role of higher education in the future of the nation.

Without a long-term idea of the country and of the role that advanced knowledge will play in the future, we will continue on in the dark, even with the best of intentions. Thus, a serious, public evaluation of the PROMEP results and of previous programs to develop the academic profession is urgently required.[6] It is absolutely necessary to generate reliable and complete information regarding academic work in Mexico. It would be very useful, not only for research work in progress but also for intelligent planning of possible reforms and development, to conduct a national census of Mexican academic staff. Fortunately, this possibility seems closer to being realized: Among new government proposals is one for the creation of the National Council for Higher Education Information, with the participation of the Instituto nacional de estadística, geografía y informática (National statistics, geography and computer science institute), the Ministry of Public Education, the ANUIES, and the Federation of Private Higher Education Institutions (*Bases para el programa,* 2000, p. 63). Such a census would, among other things, allow something that is very important: the clarification of the different kinds of professors created over the years, by disciplinary affiliation, institutional distribution, regional situation, and other factors. Ongoing surveys would enable researchers to track this important occupational group nationally and to discern the diverse forms of academic work that have evolved in different contexts and periods.

Better information on the profession would permit topics to be explored that at present can only be partially approached—such as the situation of women in academia. According to data derived from studies undertaken in 1992—a national study as well as the one carried out within the context of the Carnegie Foundation initiative—women's participation in academic work consisted of about one-third of the total and was concentrated mostly in the social and administrative sciences and the health sciences (Gil-Antón et al., 1994, pp. 219–229 and Gil-Antón, 1996, p. 322). Likewise, the national study found that of the group fully devoted to academic work, 58 percent were men and 42 percent were women, while the Carnegie study found that the numbers varied somewhat: 62 percent were men and 38 percent, women (Gil-Antón et al., 1994, p. 168 and Gil-Antón, 1996, p. 322). Another piece of information that should be followed up by more precise investigation was that women's participation is considerably higher in private institutions than in public ones. As regards formal indicators—time commitment, life-long employment, and academic degrees obtained—no major differences can be observed. However, the subject requires more detailed examination, which could show the persistence of certain limitations in women's careers, known as a "crystal ceiling," due to cultural patterns associated with women in our society. These limitations assume a break in women's careers while having and rearing children, for example. Nevertheless, compared with other professional markets, the academic profession seems to offer conditions for greater gender equality, due to the formal conditions of hiring and promotion.

In view of what has been achieved and in recognition of the soundness of having a diversity of institutions with variable purposes and missions, the country could design specific policies to address the several issues that today constitute real conundrums.

First, the employment situation of professors, as we have seen, poses a long-term problem of instability and fragmentation in terms of income. How can the academic career attract our most outstanding young students if we offer them a prospect, in the best-case scenario, after 20 years of work of having two-thirds of their income subject to periodic evaluations and uncertain budgets? Fortunately, the era is over when the same payment was given to a person who worked as to one who only pretended to work. However, it would be unfortunate if, in exchange for absolute stability and uniformity, we treated the academic profession as an aggregation of piecework.

Second, policies need to be implemented that will dignify the retirement process for the thousands of professors who built the current system. At present, these professors cannot entertain the possibility of retiring from their teaching and other academic activities, since pensions are based on their base salaries and since they would lose a precious benefit: full medical insurance coverage, precisely at an age when health problems may pose greater risks.

Third, plans for the recruitment of new professors should take regional conditions into account. In contrast to the first wave of expansion, we are today in a position to employ suitable socialization processes for the different kinds of academic activity. Furthermore, we should create recruitment, initiation, and socialization practices in academic life that are much more rational and appropriate for a higher education system more focused on teaching and learning.

Fourth, it does not seem adequate, and ANUIES states this in its prospectus, to concentrate decisions about introducing stimulus into the academic career and higher education as a whole at the central government level. In order not to repeat earlier mistakes—especially that of imposing one way of doing things on the whole country—we need to devolve the development of education largely on to the regional or state level. What may be suitable in the country's capital may not be appropriate in Chiapas or on the northern frontier. A sound federalist approach is essential.

Fifth, it is also urgently necessary to stop imposing the dominant model of the professor-researcher at all levels and in all programs and modalities of higher education in the country. Rather, the relevance of the professor-researcher should be determined based on specific levels, specialties, type of institution, and geographic location.[7] It may be necessary to classify the different curricula offered by institutions and draw up a national typology that distinguishes between the different missions of higher education organizations and relates each to the type and composition of its academic staff. ANUIES has already approved a "typology of higher education institutions," which is divided into six categories based on the type of curriculum to which the institutions give preference—ranging from those that have only two-year programs (*técnico universitario superior*) to those that concentrate exclusively on research and postgraduate programs (ANUIES, 1999, p. 17). However, institutions tend to try to climb over one or several steps in this typology—rapidly modifying their profiles—since this will give them greater prestige and access to additional funding. Therefore, we need to make an effort to reach a consensus that quality is possible in every type of institution and to provide adequate pay to all academics.

It would be very useful to take into account the proposals contained in a 1996 report of the Inter-American Development Bank that looks at the progress made by the various countries as well as the policies that have not been effective—especially the part devoted to "The Confusion between Words and the Real World" and the proposal to differentiate between educating an academic elite, the work of properly preparing professionals, the appropriate educational training of technicians, and the nature of general higher education (de Moura Castro and Levy, 1996). Each one of these functions can be properly developed, with suitable academic staff, and always with the possibility of recognizing quality and offering academic tracks adequate to the core objectives of each level (de Moura Castro and Levy, 1996, pp. 5–15). And it is time to revisit the most effective way to incorporate active professionals into academia as a link to the world of work: simply hiring professionals (cheaply) in part-time positions is no longer enough.

Last but not least among concerns are the effects on our system of the implementation of supplementary income programs over the last decade. The need to obtain higher incomes leads professors to decide how best to allocate their time and efforts; however, at the same time, institutions require their professors to work together with colleagues to implement more effective teaching and learning. However, the mere presence in a department of professors who frequently publish in international journals does not create the necessary environment for students to learn, learn how to learn, and acquire the skills and habits for collaborative work. Institutions demand—in their rhetoric—the collegial participation of academics in institutional tasks. However, they do not reward this collaborative effort; instead, they reward the performance of individual activities that enhance the specific indicators by which staff are evaluated. This contradiction is very serious, with the most affected zone being undergraduate education. During the coming years, the majority of those attending public institutions will be, as were their professors, first-generation university students. In order to achieve social mobility and a better quality of life, these students will need to obtain more from the university than a certificate.

While the impact of technology is unpredictable, the breakup of the monopoly of the university and of formal higher education as the exclusive providers of access to knowledge seems a certainty for the future. However, after years of working on these issues, a question looms large in my mind, and I wish to share it now: has the process that led to the creation of the Mexican academic profession created a socially active critical mass capable of understanding transformations in progress and of acting as an ally to society in an era that will see the opening up of knowledge? Will we know how to become true partners with students as they learn, both from us and on their own, and experience the wonders of knowledge and its applications? Or, will we act as a strong hindrance by not abandoning our traditional role, based on one-way teaching, unable to see beyond our own group interests? As Bob Dylan would say, "the answer is blowing in the wind."

NOTES

1. Big City Love (Marti, 2000).
2. It is estimated that in all of Mexico, in 1950, the number of doctoral degree holders, including refugees from the Spanish Civil War, did not exceed 100.

3. Higher education encompasses the professional associate level, university and technological undergraduate studies, the teacher training school, and graduate studies. Professional specialization, master's, and doctoral degrees are already considered as higher education matriculation by information sources. In 1960, only the university and technological undergraduate studies and graduate studies were considered. Likewise, the total number of academic positions in 1999 (192,000) corresponds to all those levels of advanced studies.

4. Here we refer only to the number of new positions added each day at the university and technological undergraduate studies level, without including graduate studies, teacher training programs, and associate professional levels, because official sources provide complete information only for very recent years.

5. The term tenure needs to be clarified. In Mexico this term describes an employment level with job stability, rather than the successful conclusion of a test period that is regulated and peer-assessed, as occurs, for instance, in the United States (Tierney and Bensimon, 1996). In Mexico it is known as *definitividad en el empleo* (having a long-term stable job). After a few months or years of continuously working—especially in those periods of greatest expansion—"tenure," or job security, was awarded, even in those cases involving part-time contracts. In Mexico a professor can have tenure with a nine-hour weekly contract, or even less. In the case of full-time contracts, tenure is obtained by means of a *concurso de oposición* (a selection process through competitive examinations), but quite often there is a candidate who has already held the position for some time. A search is opened for the post, which usually consists of a description of the incumbent's qualifications and specialization. This situation occurs mainly at public institutions, since private institutions rarely grant these types of contracts, due to the fact that they depend on the market and do not usually commit to tenured positions. In practice, however, many professors at private universities—especially elite universities—hold contracts for long periods of time.

6. Since 1970, for example, the Consejo nacional de ciencia y tecnología (CONACYT—National council of science and technology) has granted close to 100,000 scholarships for graduate studies in the country and abroad.

7. It should be noted that PROMEP recognizes the need for diversity of academic types. Likewise, diversity is taken into account in each school of higher education, but as rewards and incentives are very focused on factors related to research, professors in practice try to conform to the model to obtain more income, regardless of curricula.

REFERENCES

Altbach, Philip G. (2000). *The changing academic workplace: Comparative perspectives.* Chestnut Hill, MA: Center for International Higher Education, Boston College.

ANUIES. (1970). *Anuarios Estadísticos, 1970.* Mexico City: ANUIES.

———. (1999). *Tipología de instituciones de educación superior.* Mexico City: ANUIES.

———. (2000) *La educación superior en el siglo XXI: Líneas estratégicas de desarrollo.* Mexico City: ANUIES.

Bases para el programa sectorial de educación 2001–2006. (2000). Coordinación del Área Educativa del Equipo de Transición del Presidente Electo Vicente Fox Quesada, Mexico City.

Boltvinik, Julio. (1995). La satisfacción de las necesidades esenciales en México en los setenta y ochenta. In P. Moncayo and J. Woldenberg (Eds.), *Desarrollo, desigualdad y medio ambiente.* Mexico City: Cal y Arena.

Brunner, José Joaquín. (1987). *Universidad y Sociedad en América Latina.* Mexico City: Universidad Autónoma Metropolitana-Azcapotzalco.

Casillas-Alvarado, Miguel Angel. (2001). *La recomposition du champ universitarie au Mexique.* Unpublished doctoral thesis, Ècole des hautes études en sciences sociales, París.

de Moura Castro, Claudio, and Levy, Daniel. (1996). *Higher education in Latin America and the Caribbean: A strategy paper.* Washington, D.C.: Inter-American Development Bank.

Finkelstein, M., Seal, R., and Schuster, J. (1998). *The new academic generation: A profession in transformation.* Baltimore, MD: Johns Hopkins University Press.

Garcia-Diego, Javier. (1996). *Rudos contra científicos: La Universidad Nacional durante la Revolución mexicana.* Mexico City: El Colegio de México y UNAM.

García Salord, Susana. (1998). *Estudio socioantropológico de las clases medias urbanas en México: El capital social y el capital cultural como espacios de constitución simbólica de las clases sociales.* Unpublished doctoral thesis, UNAM.

———. (1999). Un reto para el fin de siglo: Desterrar la sombra de la duda y restituir la vigencia del interés general. Paper presented at the *Primer congreso de ciencias sociales,* Consejo Mexicano de Ciencias Sociales, Mexico City.

Gil-Antón, Manuel. (1996). The Mexican academic profession. In Philip G. Altbach (Ed.), *The international academic profession: Portraits of fourteen countries* (pp. 307–339). Princeton, NJ: Carnegie Foundation for the Advancement of Teaching.

———. (1997, December). Origen no es destino: Otra vuelta de tuerca a la diversidad del oficio académico en México. *Revista Mexicana de Investigación Educativa,* 4, 255–297.

———. (1998). Origen, conformación y crisis de los enseñadores mexicanos: Posibilidades y límites de una reforma en curso. In ANUIES, *Tres décadas de políticas de estado en la educación superior* (pp. 59–99). Mexico City: ANUIES.

———. (1999, October). El mercado de trabajo académico: Notas sobre la evolución del espacio laboral en la universidad mexicana. *Este País: tendencias y opiniones,* 103, 34–37.

Gil-Antón, Manuel. (2000a, October). Un siglo buscando doctores. *Revista de la Educación Superior,* 113, 23–42.

———. (2000b). Los académicos en los noventa: ¿Actores, sujetos, espectadores o rehenes? In Oliveira Schmidt and Alvarez Aragón (Eds.), *Entre escombros e alternativas: Ensino superior na américa latina* (pp. 155–177). Brasília, Brazil: Editora Universidade de Brasília.

Gil-Antón, Manuel et al. (1994). *Los rasgos de la diversidad: Un estudio sobre los académicos mexicanos.* Mexico City: Universidad Autónoma Metropolitana, Azcapotzalco.

Ibarra Colado, Eduardo. (1998). *La Universidad en México Hoy: Gubernamentalidad y modernización.*Unpublished doctoral thesis, UNAM.

———. (1999). Evaluación, productividad y conocimiento: Barreras institucionales al desarrollo académico. Paper presented at the *Primer congreso de ciencias sociales,* Consejo Mexicano de Ciencias Sociales, Mexico City.

Levy, Daniel C. (1995). *La educación superior y el estado en américa latina: Desafíos privados al predominio público.* Mexico City: UNAM-Miguel Ángel Porrúa.

Martí, José. (2000). *Versos.* Mexico City: Universidad Autónoma Metropolitana.

Metzger, Walter. (1987). The academic profession in the United States. In Burton Clark (Ed.), *The academic profession: National, disciplinary and institutional settings.* Berkeley: University of California Press.

PROMEP. (1997). *Programa para el mejoramiento del profesorado.* Mexico City: Secretaría de Educación Pública.

Tierney William G., and Bensimon, Estela Mara. (1996). *Promotion and tenure: Community and socialization in academe.* Albany: State University of New York Press.

Villa-Lever, Lorenza. (1996, January–March). Hacia una tipología de los académicos: Los docentes, los investigadores y los gestores. *Revista Mexicana de Sociología,* 58(1), 80–102.

3

Universities and Professors in Argentina: Changes and Challenges

Carlos Marquis

During the past several decades Argentina has experienced many political, economic, and social changes. In 1983 democracy was reinstalled after the difficult period of military dictatorship. The currency was stabilized, ending a long period of hyperinflation and initiating an economic recovery, although the number of unemployed persons did increase. As in many other places in the world, the state welfare system was dismantled during the 1990s. Since mid-2001, Argentina has again been engulfed by an economic crisis that is impacting institutions as well as a variety of social actors—including university professors, who are powerless to prevent the continuing deterioration of their working conditions.

Historically, Argentina has enjoyed a relatively high level of education compared with other countries in the region. Argentine authorities are proud of the country's schooling rate, which is comparable to that in several European countries. Argentina cannot, however, claim to have a higher education system as stable and sound as systems in Europe or the quality institutions of countries in the northern hemisphere. It should also be noted that, as in many other countries, education analysts in Argentina agree that secondary education is the sector that is actually in the more critical state.

An Overview

The rapid increase in demand for higher education in Argentina during the last quarter of the 20th century and the failure to manage this growth adequately are responsible for the "masses of professors" poorly qualified for teaching and even less so for conducting research. To worsen matters, most of the faculty managers and administrators hired by the new higher education institutions established to accommodate the rise in the number of students lack university experience.

In March 2000 a task force of experts on higher education, convened by UNESCO and the World Bank, presented their conclusions on higher education in developing countries. Their published research findings analyze higher education's current situation and future prospects. The focus on tertiary-level education was quite striking, given the fact the World Bank used to give basic education priority in its projects. A number of the issues examined by the task force have direct relevance to higher education in Argentina: university accreditation, higher education financing, university governance and autonomy, and faculty working conditions. Unfortunately, the study also found evidence that faculty "lack enough qualifications and motivation, and are poorly paid" (Task Force on Higher Education, 2000).

In Argentina the massive expansion of tertiary-level education was not accompanied by the necessary structural reforms and adequate increases in university budgets. Consequently, in spite of greater public investment in the higher education sector, the expenditure per student decreased. Simultaneously, changes were introduced into institutional management to satisfy calls for accountability. All these factors have led the Argentine higher education system into a state of crisis, which still exists despite efforts that were undertaken in order to overcome the difficulties.

Federal authorities and universities share an ambivalence toward academic faculty. While professors are considered crucial for improving the quality of higher education, the system does not provide adequate training to furnish teachers with suitable knowledge and teaching techniques. Moreover, educators suffer from low salaries and poor social and institutional recognition for their work. Paychecks for academics are far from competitive in the labor market, which limits the ability of universities to hire the most qualified educators. This has led to the deterioration of both teaching and learning standards.

The 1990s brought a number of reforms in higher education, many of which have also been implemented in other countries. Argentina introduced assessment and accreditation of higher education institutions and programs to improve quality. To foster university research activities, the authorities offered faculty economic incentives to undertake such work. There was a clear attempt to introduce transparency into academic administration, management, and budgetary matters. In the 1990s a new higher education law was passed that set new rules of the game for the sector. The country also received loans

from external banks. The World Bank, for example, helped finance
university reforms, and the International Development Bank supported
nonuniversity tertiary education.

These policy initiatives have both benefited and harmed
Argentina's higher education system—hence the words in the title of
this chapter: "changes and challenges." Unfortunately, the challenges
outweigh the actual changes.

THE STUDENT BODY

The lack of university entrance exams is partially responsible for the
large number of higher education enrollments, although most of these
students drop out soon after entering the university. Nevertheless, during the last 25 years, postsecondary education in Argentina expanded
at an annual rate of 7 percent, four times faster than the rate of increase
in the general population of less than 2 percent (Gertel, 2001). As
shown in table 3.1, around 1.5 million students attend more than
1,700 higher education institutions, which employ nearly 160,000 faculty members. According to projections, the student population in
Argentina will increase 60 percent by 2010. And this figure does not
include "new students" (mature students and students from hitherto
underrepresented sectors of society).

Rising educational and employment levels characterized much of
the 20th century. A large, educated middle class, composed mainly of
European immigrants, differentiated Argentina from other countries

Table 3.1   A profile of Argentine higher education

	Number of institutions[a]	Students		Professors N
		N^b	%	
Universities and state university institutes	43	946,506	62.2	101,339
Universities and private university institutes[d]	49	166,181	10.9	17,754[c]
Higher education nonuniversity institutes	1,664	409,511	26.9	40,160[e]
Total	1,756	1,522,534	100	159,253

Notes
[a] From Ministry of Education, 2000.
[b] From INDEC 2000. (1998 data).
[c] Author's estimates.
[d] Nonuniversity higher education institutes are 70 percent public and 30 percent private.
Public institutions serve 60 percent of students and private, 40 percent.
[e] From Giuliodori and Nychaszula, 1998.

in South America. However, structural economic difficulties as well as the current economic crisis and increasing unemployment (currently 17 percent of the working-age population is unemployed) are adversely affecting the younger generation and impoverishing the whole country. Graduates of higher education institutions are not exempted: the unemployment rate among those in the population with a higher education diploma rose sharply from 1.1 percent in 1990 to 7.8 percent in 2001 (Gómez, 2001).

An interesting development is the increasing demand for tertiary-level education at nonuniversity institutions, which now enroll 27 percent of students. This trend is fostered by the current government as a strategy to enlarge education coverage and differentiate the educational offerings (Sánchez Martínez, 2001).

In Argentina's higher education system, undergraduate studies, or *licenciaturas*, are referred to as graduate studies and master's and doctoral programs are referred to as postgraduate studies. The *licenciatura* consists of a higher level of study than for the American bachelor's degree but a level less advanced than for the American master's degree. *Licenciatura* programs entail five years of study. Students rarely complete the degree within this period, taking an average of seven years to attain their first degree. Argentines, therefore, typically receive their first diploma at the age students in developed countries obtain their second diploma. Recent studies show that the graduation rate is 57 percent (Landi and Giuliodori, 2001). This is a respectable figure, but a mere 20 percent of students graduate on schedule.

With regard to socioeconomic background, most students come from higher-income families. However, new studies show that students from poorly educated, low-income families now constitute 20 percent of the total university population (García de Fanelli, 2001). It should also be noted that 28 percent of the student population is over 25 years of age due to the promotion of continuing education and the increasing demand for postgraduate studies. The presence of so many older students is also a reflection of the many years it can take students to complete their degrees.

While the high number of educators who work part time is characteristic of Argentine universities, many students are enrolled on a part-time basis as well. Approximately 40 percent of students in Argentina work, which means that only 60 percent are full-time students. This constitutes a major difference between Argentine and European students, as 90 percent of the latter are enrolled in full-time programs (Gertel, 2001).

Actually, the 60 percent of students who do not work are not necessarily full-time students. This is because most program schedules are not designed to keep students engaged full time. As a result, few educators or students are in residence full time at higher education institutions, although the figures vary according to program. Thus, only 12 percent of students in medicine have jobs, while 50 percent of liberal arts students do have jobs. A student's age and gender are also factors: The older the student the more likely it is that he or she works. In addition, 45 percent of male students support themselves whereas only 35 percent of women do so. Thus, it might be easier to describe students as *studying workers* than as working students. This is even more the case with respect to postgraduate students.

HIGHER EDUCATION INSTITUTIONS

The first higher education institution in Argentina was Córdoba University, founded in 1613. It was also the birthplace of the 1918 Cordoba Reform, which contains the old principles still in force at universities in Argentina and other Latin American countries. It was also the first state university to introduce tuition fees—a practice rejected by other institutions and one that provoked many conflicts between government officials and university managers.

The University of Buenos Aires (UBA) celebrated its 180th anniversary in 2001. With a student population of close to 250,000, it has earned the status of a *megauniversity*. As one might expect, UBA's organization, administration, and management are highly complex. Later in this chapter UBA's main characteristics as an academic workplace will be examined. The many new state and private universities round out the general picture of the 90 universities in Argentina's higher education system.

In total, there are 1,700 higher education institutions, of various kinds, in Argentina: state and private, religious and secular, small and large, old and new, and with diverse programs. Their forms of administration differ tremendously. Some of these institutions carry out teaching, research, and knowledge transfer or outreach, while others offer only instruction. Of course, institutions vary in terms of the quality of their programs and faculty.

Most Argentine higher education institutions follow the old Napoleonic model, which is centered on professional training and organized around faculties, schools, and chairs (*cátedras*), with limited research and knowledge transfer activities. At other higher education institutions, modernization took place after the 1960s, leading to departmental organization, and centered on research and related activities. Institutions responded in different ways to the reforms of the 1990s and the introduction of assessment, incentives, and other procedures. Some institutions welcomed the changes, others simply went along with them, and many vehemently rejected them. No doubt the reforms have had an impact on academics and have generated new practices regarding evaluation of teaching performance, incentives, and assessment. The reforms have also affected faculty relationships with their institutions; faculty have had to respond to the new management models.

FUNCTIONS AND VARIABLES

In their recent study of higher education policy in Latin America, Levy and Moura Castro (2000) identify four different functions of higher education institutions: academic leadership, professional training, technological training, and general education. The differentiation of functions may occur not only across different types of institutions— such as research universities, schools and colleges, technical institutes, or community colleges—but also within the same institution. Yet, unfortunately, all types of faculty within an institution receive the same or similar treatment—even when faculty and administrators explicitly wish to establish differences. It is very important to differentiate institutional structures, projects, and possibilities to decide the best policies for each. In its work, the Task Force on Higher Education recommended an explicit stratification to allow each institution to grow according to its strength and to serve different needs (Task Force on Higher Education, 2000).

Another way to understand differences in Argentine higher education can be found in the variables proposed by Clark (1983) and Brunner and Martínez Nogueira (1999): first, the tendency of the institution, traditional or innovative; and second, the type of program, academic or professional. In combination, these variables can assist in forming innovations in the system that may not be obvious in the formal institutional organization. Innovation and adaptation may occur in both traditional and innovative institutions. Likewise, both discipline-based academic programs and professional programs may draw on personnel from within or outside.

RANK, WORKLOAD, AND REMUNERATION

In most cases, public and private universities classify their faculty into categories—professors and lecturers—with three ranks. For professors the ranks are full (*titular*) professor, associate professor, and assistant professor. The three ranks of lecturers can be referred to as first, second, and third.

Universities determine a faculty member's rank on the basis of an individual's qualifications, experience, and expertise. Each university values different credentials, but full professors and chairs have published relevant studies, led research teams, and enjoy academic leadership in their disciplines. To illustrate the way the system operates, a degree program in sociology will include many subjects, called chairs, each of which is normally taught by a full professor. Thus, a sociology program with 28 chairs would be expected to have 28 full professors, each one responsible for his or her team.

While each university establishes its own recruitment criteria, associate professors are generally required to have almost the same credentials as *titulares,* which makes the replacement of *titulares* easier. Assistant professors are generally younger and less experienced than other faculty members. Entry-level positions to a university career are held by lecturers.

The pyramidal shape of the chairs allows for the promotion of faculty members only when there is a vacancy, regardless of their qualifications. Promotion to a higher rank is not possible unless an opening exists for an appropriate position on the career ladder.

Faculty weekly workload, called *dedicación,* depends on the type of contract: Full time stipulates 40 hours; part time, 20 hours; and hourly contracts, usually 10 hours. However, because salaries are generally poor, contract terms are frequently not enforced, thus allowing faculty to work fewer hours than specified. This results in the deterioration of academic performance.

Table 3.2 illustrates the distribution of faculty in the whole state university system by rank and workload. It is important to point out that only 3.3 percent of 101,339 university positions are held by those with the highest rank and workload. Some universities have created a new category for those faculty members who work on full-time contracts exclusively for the university and so receive a special bonus. However, the most innovative trend consists of hiring professors for a limited period of time, usually for a semester. These contracts are used to recruit educators from other universities to perform specific kinds of activities, such as teaching. It is mainly the new universities that have adopted this innovative trend.

Faculty remuneration depends on four factors: academic rank—*titular,* associate, assistant, or lecturer; workload; seniority; and various types of incentives. The maximum bonus for seniority actually exceeds the base salary. The incentives are offered by the federal

Table 3.2 Faculty at public universities, with rank and workload, 1998

Rank	Full-time	Part-time	Hourly	Total
Full professors	3,298	3,892	7,911	15,101
Associate professors	1,373	937	2,040	4,350
Assistant professors	3,782	5,421	13,223	22,426
Subtotal	8,453	10,250	23,174	41,877
Lecturers 1st	3,786	8,135	16,385	28,306
Lecturers 2nd	1,699	4,084	16,243	22,026
Lecturers 3rd	0	0	9,130	9,130
Subtotal	5,485	12,219	41,758	59,462
Total	13,938	22,469	64,932	101,339

Note: Data from Ministry of Education, 2000.

government to faculty members who adopt certain guidelines in their research. Other type of incentives are offered directly by institutions.

A large number of faculty members volunteer their time without remuneration. Third-rank lecturers are not the only academics who work for free. Nearly 25 percent of academics volunteer their service, and they do so for the honor associated with being a faculty member. Volunteering at a university also means being deprived of social security benefits. In many cases, faculty hold paid positions with minimum workloads, despite their poor salaries, to obtain social security benefits.

PROFILE OF THE PROFESSORIATE

As of 1998, there were 101,339 academic positions in Argentine state universities. Professors comprised 41 percent of this number and lecturers, 59 percent. Only 14 percent of faculty members were on full-time contracts, while 86 percent were on part-time contracts or worked on an hourly basis.

Table 3.2 shows the distribution of the 101,339 faculty positions among 77,732 people—a phenomenon explained by the fact that many people hold several part-time or hourly positions. Faculty frequently sign three or four contracts, each for 20 hours a week. If followed to the letter, these contracts would be impossible to fulfill, because it would require a person to work an average of 60 to 80 hours a week. This illegal practice is called *incompatibility,* and will be discussed later in this chapter.

Table 3.3 shows that women outnumber men in full-time positions. It is quite probable that the low salaries of professors are keeping an increasing number of men away from the higher education system.

Table 3.3   Faculty workload by gender,
1998 (percentages)

Faculty	Men	Women
Full-time	48	52
Part-time	59	41
Total (101,339)	55	45

Note: Data from Ministry of Education, 2000.

Age distribution is spread evenly: 13 percent of faculty are under age 30; 30 percent between ages 30 and 39; 28 percent between 40 and 49; and 27 percent over age 50 (Ministerio de Educación, 2000). A distinguishing trait of the system, as shown in table 3.2, is that only 20 percent of professors and 9 percent of lecturers hold full-time positions. Full-time, full professors hold only 3,298 of all posts—an alarmingly low 3.3 percent of the total.

SKETCH OF BUENOS AIRES UNIVERSITY

The University of Buenos Aires is 180 years old and enjoys a reputation as the most important institution of its kind. UBA has a population of 253,260 undergraduate and 8,809 graduate students. A total of 24,508 faculty members work for this university. The university is traditionally organized around faculties and research institutes. In addition, UBA runs two high schools.

Because of its large size and complex organization, several university "patterns" coexist within UBA, though the dominant format is found in its professional programs—such as medicine, law, accountancy, and engineering. The dominant institutional pattern for organizing research can be seen in the exact and natural sciences, which offer programs in many disciplines—including mathematics, physics, chemistry, and biology.

Career-oriented programs focus on student development, and the professors are mainly concerned with teaching. These professors generally spend just a few hours a week at the university and lack suitable facilities to conduct their work activities. By contrast, the principal focus of research programs is on knowledge creation and transfer. UBA research institutes are among the best in the country, with highly qualified researchers in well-equipped environments devoting themselves to academic life.

Table 3.4 Faculty at Buenos Aires University, by workload, remuneration, and rank, 2000

	Remunerated professors	Remunerated lecturers	Volunteer faculty	Total[a]	
				N	%
Full-time	1,190	1,175	na[b]	2,365	10.2
Part-time	1,324	1,552	na	2,876	12.4
Per hour	3,644	7,750	6,605	17,999	77.4
Total	6,158	10,477	6,605	23,240	100.0
	26.5%	45.1%	28.4%		

Notes: Data from Faculty Census 2000, Buenos Aires University.
[a] The 974 teachers from the high schools affiliated with Buenos Aires University are not included in this table.
[b] na=not applicable.

It is interesting to note that 28.4 percent of UBA faculty work without payment. A surprising 66 percent of faculty volunteer at the School of Medicine. In the School of Law, as well as in the Economics Department, more than 33 percent of academic personnel are not remunerated. On the other hand, only 1.2 percent of staff affiliated with mathematics, physics, chemistry, and biology programs volunteer their services. In liberal arts departments, 6.1 percent teach on a voluntary basis.

Distribution of full-time professors follows the same pattern. In the exact sciences, natural sciences, and agronomy more than two-thirds of educators work full time. In law, architecture, and economics less than 5 percent of academic staff work full time. The percentage of professors working full time does not correlate with the number of students in the program. Approximately 16,000 students are enrolled in the exact sciences, natural sciences, pharmacology, biochemistry, agronomy, and veterinary science. Those studying law, economics, medicine, and architecture total about 109,000. In short, disparate academic units coexist within UBA, resulting in widely differing conditions for educators.

Governance and management are particularly difficult at a university as large as UBA. There have been many attempts to reform the system according to the French higher education model—such as splitting the university into UBA 1, UBA 2, and so on—but the idea did not succeed. UBA also has a history of conflicts with the federal government. Though it is a top-ranking university in several fields, UBA is opposed to the quality assessment and accreditation system contained in the 1995 higher education law. In fact, UBA has even appealed the law in court.

Over the past ten years, several public and private universities of varying quality have been founded on the outskirts of Buenos Aires City to offer alternatives to UBA. These new and well-ranked institutions recruited many of their professors from UBA, tempting them with better academic prospects and salaries. The new institutions are also able to offer more flexible working conditions, which in many cases represents an improvement over UBA's strict organization. Still, UBA enjoys high prestige due to its quality and long-standing reputation.

THE INCOMPATIBILITY ISSUE AND TAXICAB PROFESSORS

As mentioned earlier, the explanation for having 77,732 people in 101,339 academic positions in Argentine universities is that many of these individuals hold more than one job. Many professors thus carry an excessive workload, despite the specific and clear regulation that limits faculty to working 40 hours per week, except under special circumstances.

Educators frequently sign on for more hours per week than they can practicably carry out. This is the "incompatibility" issue referred to earlier. Professors who work at multiple institutions must limit themselves to just conducting classes at different campuses. This is the phenomenon of the "taxicab professor," which has become a popular alternative to obtaining a full-time position. Still, some faculty do not hold multiple jobs at different universities—they manage to accumulate an incompatible workload at just one institution.

Another common way faculty manage to hold incompatible positions consists of working full time at one university while holding another job somewhere else. Periodically, federal and university authorities punish faculty who practice incompatibility, but most of the people involved disregard the regulations with impunity.

In addition to the low salaries, the rigid university contracts negotiated with the faculty unions have created the incompatibility phenomenon, which adversely affects the quality of education. The issue of incompatibility is at the core of the current disputes between the faculty unions and government authorities. The unions say that incompatible positions are the only way educators can make ends meet.

There have been some creative attempts to resolve the issue of incompatibility. Litoral National University, which was founded in 1919 and now has 20,000 students and 3,300 professors, formed a Directive Council (which included professors among its members) to review the incompatibility issue; after much study and deliberation, the council proposed a new integrated system. This system includes three general categories of employment: simple, with 10 hours a week; part time, with 20 hours a week; and full time, with 40 hours a week. Contracts for full-time faculty who work exclusively for the university were further split into three subcategories according to their respective weekly workload: full-time *A*, 48 hours; full-time *B*, 43 hours; and full-time *C*, 40 hours. People holding two positions entailing 20 hours each would be included in this last category. Those belonging to the full-time *A* category would be prohibited from having any other job, and would be compensated with an incentive. Those falling into the full-time *B* category would suffer a 20 percent reduction in their salaries but would be allowed to work up to 12 extra hours a week off campus. The last provision aroused major conflict because of the salary reduction.

Private Universities

The first private universities were established in the second half of the 20th century, bringing to a close the established tradition of public, free, and secular education. By 1958 there was great public debate surrounding secular versus religious education and free versus paid education. Finally, after passing many regulations, the government authorized the establishment of private universities. However, not until the 1990s were these regulations properly implemented.

The first private universities were Roman Catholic institutions, while later institutions were affiliated with the corporate sector. Because of the vastly disparate quality of private universities, for several years no new private institutions were authorized. Beginning in the 1990s, quality institutions were able to win approval from the University Assessment and Accreditation Federal Committee (CONEAU). Now, private universities are successfully competing in the area of career-oriented teaching. Study programs in such fields as business, engineering, and medicine have found a new quality alternative in the private sector.

Nevertheless, a sharp divide still exists between public and private universities. Private universities are not supported with public resources. Although they are allowed to compete for government research funding, scientific research is still concentrated in the Scientific and Technological Research Federal Council (CONICET), at public universities, and at a very small number of private universities. In general, private institutions are financed by fees collected from their students; rarely can they count on receiving supplementary funds.

Only limited data are available on professors working at private universities. However, analysis of several of these institutions demonstrates that private universities generally recruit their personnel from public universities. The percentage of full-time staff at private universities is around 5 percent, which corresponds to the situation at UBA in similar professional programs.

Part-time professors work at both state and private universities. Private university salaries are generally better than those at state institutions. At the private institutions, higher-ranking faculty receive $20 per hour, making their average monthly income $200 to $400, depending on their workload. These instructors are paid only for the classes they actually teach.

There are some innovative private universities that are determined to excel. They charge high fees and also receive financial aid from wealthy corporations. They offer very competitive packages similar to those in the private labor market. Faculty hired by these elite private universities are likely to earn $5,000 a month.

UNIVERSITY FINANCING

Although public universities are scattered throughout the country, they receive all their funding from the federal government rather than from local or provincial governments. Negotiations over grants are always difficult, and each year the petitions of some institutions are thwarted when their representatives seek approval from Congress for inclusion in the federal budget, setting the stage for much conflict. There is an ongoing struggle between the Ministry of Education and the Ministry of the Economy—the former demanding more resources, the latter offering not enough. Another battle is being waged by members of Congress, who annually renew the confrontations over appropriations, fighting to get bigger grants from the federal government for the provinces they represent. However, the main conflict involves the Ministry of Education and the university rectors, who also serve at the same time as university representatives, political party members, and advocates for their provinces. Rectors put all their energies into getting more resources, a task for which they are held to account by their respective communities.

Since the 1990s several governments, of differing political parties, have initiated negotiations with the universities over new budget formulas. It is the government's wish to provide funds based on rational considerations instead of political skill or confrontation. Committees comprised of representatives from the Ministry of Education and the universities were formed to draw up mathematical formulas covering all the possible variables. The parties arrived at a reasonable formula regarding budget allocation to end the political struggles that previously controlled the process. To avoid the risk that any university's share in the general budget would be cut, it was agreed that new criteria would be adopted only in conjunction with a budget increase.

Just as expectations rose for an equitable distribution of funds, however, the budget was frozen, and may even be cut. The committee's formulas were reduced to a mere intellectual exercise with no positive result. Under these circumstances, in 2000, the minister of education made an attempt to allocate only 20 percent of the total budget for the universities. But the controversy over the minister's maneuver, together with other political problems, led to his resignation.

In 1999, 2000, and 2001 the annual budget for public universities was $1.8 million, equivalent to 0.65 of GDP. Between 1993 and 1999 state financing rose 70 percent, but since 1999 this figure has not increased; now, funding may even be reduced. The government that came into office in 1999 put all its efforts into eradicating the fiscal deficit, and at the end of 2000 it decided to freeze public expenditures for the next five years (2000–2005). The immediate consequence was the loss of any possibility of an increase in state resources for universities. Currently, Argentina is facing a new economic and social crisis, which has led the government to shrink public expenditures, including allowances for state universities. The big question is how to keep universities operating.

Public universities have generated resources of their own, amounting to about 10 percent of the funding from the federal government. Between 1993 and 1998 these self-generated resources increased by 80 percent, not including fees collected for graduate studies. During the 1990s, state universities managed to augment their financial resources, both from the federal treasury and through their own efforts.

In Argentina it is unacceptable to discuss alternative ways of financing public universities. In this country, the concept of education is associated with the terms public, free, and secular. It seems impossible that any political party hoping to win office would campaign against this long tradition, even if party members privately deemed reform necessary.

Argentina faces a serious dilemma—how to finance higher education given the many challenges, obstacles, and constraints. This crisis of declining federal support comes at a time of increased enrollments in tertiary education. It will be difficult to increase the universities' own resources in a stagnant economy and in the absence of a tradition of linkages between the productive sector and higher education. Moreover, there has been a clear refusal on the part of several sectors within the university community to participate in the financing of the system. Finally, the sustained opposition of political parties to introducing fees for undergraduate studies removes the option of that method of funding.

The Argentine economy is at serious risk of collapsing due to the country's failure to repay external loans and their services. The government and most economists think that if a default takes place, the country will be further weakened for a long time to come. Global financial analysts who measure a country's likelihood to default on loans have classified Argentina as a country at risk. In fact, since the second half of 2001, Argentina has stood second in the world, after Nigeria, among the most high-risk countries in which to invest.

The critical situation Argentina currently faces is the direct consequence of the increasing amount of external loans that the country procured over the last three decades. This situation has worsened during the past three years, owing to an abrupt decrease in productivity and revenue.

To deal with this serious situation, the federal government has adopted a strategy of zero deficit—that is, spending only what is collected through taxes, without taking out new loans, which would likely be impossible to get from the financial markets and would in any case carry huge interest rates. The government is determined to avoid additional debt and to guarantee payment to external creditors in spite of the impossibility of increasing revenue under current economic conditions. To accomplish these objectives, federal authorities are trying to decrease expenditures by reducing federal salaries and pensions by 13 percent, among other restrictive measures. It is still unclear what the outcome of these initiatives will be.

It is interesting to note the response to these measures at public universities, where employees, like all those in the federal sector, have opposed a salary reduction. Once the law was passed, the government reduced its budget allocation to universities by 13 percent. However, being autonomous, many universities, including UBA, decided to cut back on other expenses to avoid reducing salaries. This approach makes university personnel privileged in comparison with other federal workers but simultaneously deprives institutions of the ability to use funds for investment and maintenance. Paradoxically, researchers with CONICET have suffered a salary reduction, while researchers at universities have not. Nonetheless, students and professors are clearly opposed to the government's official policy decisions.

FACULTY REMUNERATION
Traditional Universities

The economic circumstances described above are a factor in the low salaries of faculty and university administrators, but this is not a new phenomenon. Poor compensation existed before this crisis, despite the fact that public universities allocate more than 80 percent of their budgets to salaries. This leaves too little for other expenses, especially for investments. In 1997 universities allocated 81.4 percent of their combined budgets to pay the salaries of 149,000 faculty and administrators, which translated into an average monthly salary of $700—not enough to make ends meet in Argentina, which has a high cost of living.

The current pay scale at traditional universities consists of the base salary for each faculty category, plus seniority compensation—which can be even greater than the salary and is related not to performance, but simply to years of experience. Some universities also give supplementary incentives for instructors who hold a diploma. Table 3.5 shows the extremes in remuneration at a traditional university.

Innovative Universities

The 1995 higher education law changed the way salary compensation for professors was determined in Argentina. Previously, the government was responsible for establishing a scale of remuneration for the various professorial levels. Today, however, the process is more decentralized, as universities are allowed to negotiate salaries more freely with faculty. Traditional universities tend to remain stuck in the rigid

Table 3.5 Faculty remuneration, by rank and workoad, 2001

	Per hour		Full-time	
	0 years experience	Maximum seniority	0 years experience	Maximum seniority
Full professor	171	256	1,041	2,257
Associate professor	161	242	980	2,126
Assistant professor	138	210	849	1,842
Project leader	120	184	743	1,614
Project leader assistant	107	167	668	1,452

Source: Data from Mar del Plata State University.
Note: Most public universities share the same remuneration pattern. Since 1995 there has been a monetary incentive, under certain regulations, for professors who conduct research. This program benefits 18,000 out of 77,000 (23.4 percent) faculty members.

structures of the past, while newer universities with more flexible regulations are benefiting from decentralization.

One such institution is State Quilmes University, which was established in 1989 and is known for the high quality of its academic offerings and organization, as well as its mission of conducting applied research and knowledge transference. This university has carried out an important threefold reform in its remuneration system, improving the level and composition of salaries, faculty administration, and organization—so much so that many other new institutions want to emulate its approach.

The reform at State Quilmes University has focused on three main aspects: (1) the criteria for hiring, training, and administering academic staff; (2) a salary scale with basic elements that differ from most Argentine state universities; and (3) development of continuous assessment mechanisms for promotion, hierarchy alterations, and remuneration. There are three possible weekly workloads: 40 hours, 30 hours, and 20 hours. No 10-hour weekly contracts exist. Schedule control is maintained through electronic cards that professors now use to clock in and clock out, a practice unique among Argentine higher education institutions.

One of the central aspects of this innovative system is the incorporation of variables related to efficiency and productivity. Faculty salaries are made up of a base salary that varies along a scale with 15 gradations, which allows greater flexibility than the five grades used at most public universities; compensation for experience; compensation for academic merit; technical assistance for projects; and compensations for management responsibility.

In this way, a person entering the system as an entry-level faculty member or researcher can move to a higher rank every two years, with the possibility of reaching the top rank in 28 years. The entry-level monthly salary is $900, and provided the individual receives all possible incentives by the time he or she is a full professor, the total monthly salary could reach $4,800—more than double the earnings of a full professor who has worked exclusively and attained maximum seniority at a traditional university.

UNIVERSITY POLITICS

Universities played an important role in promoting democracy and resisting subjugation by military dictatorships. Most members of the Argentine university community were proud that they were able to defend their autonomy and to ensure joint governance between professors and students, free tuition, open public contests for faculty entering the system, and secular teaching.

Public universities are autonomous and set their own regulations, except those already established by law. They select their own leaders and administer themselves. Their form of governance resembles that of collegial bodies, but their electoral disputes recall those in the political environment. At public universities faculty and students may participate in university management in two contexts: in institutional governance through collegial boards and through unions. Professors make up at least 50 percent of the membership of governance bodies; students must fulfill certain qualifications in order to participate (e.g., they must be senior-level students).

The unions defend the faculty members' interests and the student organizations serve the students' interests, in addition to becoming involved on a wide range of political and social issues. Participation in university governance is determined through elections, in which members of the university community select candidates from lists of students and professors. During the period in which democracy was seldom practiced in society, university elections became quite popular and were followed with interest by the public. For a long time developing political competence within the university was considered excellent training for eventual participation in a democratic society and public life. Many student leaders later went on to become members of Congress or to hold other political offices.

Universities once enjoyed a reputation for transparency when selecting faculty and leaders, honesty in their fiscal management, and efficiency in the administration of their institutions. Higher education institutions expressed solidarity with the poorest sectors of society, providing extension programs and certain forms of social service. Nonetheless, over time, universities lost these positive features. They could no longer guarantee their students economic success and they lost their social-minded spirit when democracy returned to Argentina and political activities began to revolve around the political parties and other social organizations.

Although they retained the same collegial form, universities were changing their power structure. No longer were important decisions being made by academics on collegial boards. The growing complexity of the institutions placed administrators in the main leadership roles. Administrators and managers introduced bureaucratic logic within their administrative hierarchies, keeping institutional power for themselves.

Moving from an academic to a bureaucratic style would not have proven so traumatic if reforms in government structures had kept pace. In other countries, academic decisions are kept separate from administrative matters because the two areas have different rationales and aims. In Argentina, the spheres were mixed, which has led to weakened decisions in all areas. This situation has favored neither administrators nor academics but, instead, political parties, whose influence in the university sphere has increased.

As democracy receded in the universities, groups with close ties to the political parties managed to strengthen their power at public universities. The presence of political parties has produced a stifling atmosphere at universities, which are now subject to the same corruption and authoritarianism that exist in the society at large.

THE CONTEST

Argentina's higher education law establishes several rights for professors: to embark on an academic career by means of a public and open contest, in which one's credentials are compared with those of other applicants; participate in the management of one's own university; advance one's academic career; and participate in union activities. Faculty obligations are to follow the rules that regulate the institution to which they belong, responsibly participate in academic activities, and develop their professional skills according to the demands of their academic field.

Only those professors who obtain their positions through an open contest are allowed to vote or run for office in conjunction with the governance of their respective institutions. To ensure that governance remains democratic, the law establishes that 70 percent of a university's total faculty must obtain their posts through contests. The contest is a strategic issue in the Argentine university system because it legitimizes access to the institution and guarantees professors at least seven years of stable employment and at the same time is the way for educators to attain what is known as "citizenship in the university republic."

Nevertheless, on many occasions, the open contest becomes a charade, subject to manipulation on the part of both evaluators and competitors. Contests should be the most transparent way of recruiting into higher education people who are returning to Argentina from abroad after earning their doctorates or other advanced degrees. However, these candidates are frequently rejected because they are perceived as a threat by other professors who may not have kept up to date in their fields. Moreover, it is rare for a professor who has once won a contest to fail the next one seven years later. This means that a professor's credentials are not properly verified, as prescribed in the university reform of 1918.

Access to a professorship confers a certain amount of job stability. However, although the law is clear with respect to access, it is not so clear about stability. Consequently, each university makes its own decisions regarding a professor's appointment, and procedures differ from one institution to another. At traditional universities, after a successful contest professors enjoy seven years of stability until they must participate in another contest. Innovative universities, however, allow professors to achieve permanent job stability (similar to tenure in the American system), provided their annual or biannual assessments are positive. If an assessment is negative, the system allows for some kind of action to be taken against the faculty member in question.

Undoubtedly, the more typical way of hiring professors—increasingly common owing to the current crisis in higher education—is on semester-long or annual contracts. Unfortunately, this practice provides faculty with neither stability nor the right to participate in institutional governance. Naturally, the university unions strongly oppose this hiring system, in spite of the fact that this is the only method that allows the younger generations of professors access to the academic profession.

CONCLUSION

There is definitely a market for professors, generated by the increasing demand for higher education. This market is differentiated by academic workplace, discipline, and workload.

Relatively few professors can be identified as typical full-time university employees. Many faculty members work professionally in several academic positions at more than one institution, which prevents them from actually "belonging" to any one institution. At the same time, there are many professionals who pursue full-time careers and teach part time at a university; generally, these individuals have a tenuous relationship with the universities at which they teach.

In addition, a high percentage of professors work without pay at public universities, and a sizable number hold positions that carry minimum workloads and meager salaries. In fact, these two groups are propping up the higher education system in Argentina. It is important to point out that many of these professors are working toward a master's or doctoral degree without any scholarship assistance.

Faculty working at traditional universities perform their activities in very uncomfortable settings. They do not have offices or facilities, except for the symbolic professors' lounge. Traditional universities have facilities only for those who work in a laboratory, though some institutions are trying to remedy this situation. Overall, however, the "taxicab professor" phenomenon is not only tolerated but is even encouraged by universities.

Entry into university careers, job stability, and promotion of professors are controversial issues that are sometimes dealt with ineffectively. The open contest is supposed to assure candidates that they are going to be judged by a qualified jury focused on selecting the best possible professor. In reality, the process is encumbered by the authorities' manipulations and the resistance of faculty who fear that their own competence might be brought into question by certain candidates in the contest. Access to an academic career is not exempt from a certain arbitrariness on the part of those who make the decisions.

The importance given to political party affiliation in Argentina is particularly striking. In other countries, conflicts may be of a theoretical, professional, disciplinary, or organizational nature. In Argentina, conflicts are almost exclusively between members of different political parties, or even between people with different ideological tendencies within a party. The pervasiveness of partisan politics has a definite impact on higher education governance and policy. It should also be noted that student organizations have a voice in the election of institutional leaders and that they, too, can be highly politically oriented.

In Argentina universities are still debating over the best course to follow. Although the innovations of the last decade have been incorporated, they did not impact all universities in the same way. These innovations were adopted by a devastated, poorly designed and even more poorly administered system. Not only does this system require more resources, it also needs more reforms and some radical changes.

In different ways, all professors suffered as a result of the politics of the last decade, particularly because of the new rules necessitating that professors improve their skills. Now, faculty are pursuing postgraduate studies en masse. A small proportion of faculty have welcomed the modernizing policies—the academic professors who work full time, have Ph.D.s, and are dedicated to both teaching and research. Most faculty, however, belong to traditional institutions where full-time work is rare. Among these faculty, the new requirements were poorly received.

Two trends that have been identified as characterizing the academic profession globally (Altbach, 2000) also apply to Argentina—the increasing number of faculty looking for part-time positions and the expansion of full-time, limited contracts without job security. Argentine higher education is an exception in some ways, as it is looking to increase the percentage of full-time faculty. At the same time, Argentina's universities hope to adopt more flexible contracts. The challenges that Argentine universities will face in the coming decade are numerous. Serving the increasing demand for higher education will require a more complex and diversified system, with more complex organization, governance, and administration. These innovations will be very difficult to achieve. Finally, universities must grapple with the perennial problem of financing, which is not only a question of funds and resources, but also of winning support from a variety of sectors in the broader society, whose support will be needed (Marquis, 2001).

References

Altbach, P. G. (2000). The deterioration of the academic state: International patterns of academic work. In P. G. Altbach (Ed.), *The changing academic workplace: Comparative perspectives.* Chestnut Hill, MA: Center for International Higher Education, Boston College.

Brunner, J. J., and Martínez Nogueira, R. (1999). Evaluación externa del FOMEC. *Infomec* no. 8, 15–52. Buenos Aires: Ministerio de Educación.

Clark, B. (1983). El sistema de educación superior: Una visión comparativa de la organización académica. Mexico: Universidad Futura.

García de Fanelli, A. (2001). Los estudiantes universitarios en la Argentina. In A. Jozami and Sánchez Martínez (Eds.), *Estudiantes y profesionales en la Argentina.* Buenos Aires: EDUNTREF.

Gertel H. (2001). Los estudiantes de la educación superior en la Argentina. In A. Jozami and E. Sánchez Martínez (Eds.), *Estudiantes y profesionales en la Argentina.* Buenos Aires: EDUNTREF.

Giuliodori R., and Nychaszula, S. (1998). La enseñanza de nivel superior no universitaria de formación técnico profesional. In Delfino et al. (Eds.), *La educación superior técnica no universitaria.* Buenos Aires: Nuevas Tendencias.

Gómez, M. (2001). Mercado de trabajo e inserción laboral de los profesionales universitarios. In A. Jozami and E. Sánchez Martínez (Eds.), *Estudiantes y profesionales en la Argentina.* Buenos Aires: EDUNTREF.

Instituto Nacional de Estadísticas y Censos (INDEC). (2000). *Anuario estadístico de la República Argentina.* Buenos Aires: Ministerio de Economía.

Landi, J., and Giuuliodori, R. (2001). Graduación y deserción en las Universidades Nacionales. In A. Jozami and E. Sánchez Martínez (Eds.), *Estudiantes y profesionales en la Argentina.* Buenos Aires: EDUNTREF.

Levy, D., and Moura Castro, C. (2000). *Myth, Reality, and Reform.* Washington, DC: Inter-American Development Bank/Johns Hopkins University Press.

Marquis, C. (2001). El financiamiento universitario en la Argentina. *Revista de la Educación Superior,* no. 117, 93–97.

Ministerio de Educación. (2000). *Anuario de Estadísticas Universitarias 1998.* Buenos Aires: Secretaría de Educación Superior.

Sánchez Martínez, E. (2001). Nuevos datos sobre la educación superior. In A. Jozami and E. Sánchez Martínez (Eds.), *Estudiantes y profesionales en la Argentina.* Buenos Aires: EDUNTREF.

The Task Force on Higher Education and Society. (2000). *Higher Education in Developing Countries, Peril and Promise.* Washington, D.C.: World Bank.

4

THE CHANGING ACADEMIC
WORKPLACE IN BRAZIL

*Elizabeth Balbachevsky and
Maria da Conceição Quinteiro*

The last decade of the 20th century was a decade of great changes in Brazilian society. Until the 1990s, Brazil had a closed economy. The prevailing notion among the Brazilian elite linked development to the government's success in protecting Brazilian enterprises. Autarchic, hierarchic, and centralized perspectives predominated.

In the 1990s, the opening up of the economy, even though to a moderate degree, exposed Brazilian enterprises to hitherto unknown levels of competition. Monetary stabilization,[1] a successful privatization program, and a new regulatory framework adopted by the constitutional amendments of the 1990s have created a new macroeconomic environment. These changes have put new pressures on the Brazilian higher education system, especially on the demand side. This chapter will examine the evolution of Brazilian higher education under the new framework created by the reforms, with special emphasis on the changes in the academic workplace.

A BRIEF HISTORY

It is usually said that Brazil is a latecomer in the history of higher education. In fact, during the greater part of Brazil's colonial history, Portuguese laws and ordinances prevented the establishment of higher education institutions. It was only in the early 19th century, when the Portuguese royal family fled to Brazil to escape from Napoleon, that the first higher education institutions were founded. They started as professional schools, mostly in medicine, engineering, and law.

The separate, nonuniversity professional school was the only institutional model until the 20th century. Not until the early 1920s was establishing a university seriously considered by the Brazilian authorities and elite. The first university, the Universidade do Rio de Janeiro, was founded in 1920, and the first university law was enacted in 1931.

The years between 1930 and 1950 were a time of growth and diversification in Brazilian higher education. In 1934, the state of São Paulo, with the country's most dynamic economy, inaugurated its own university, the Universidade de São Paulo. Between 1930 and 1945 other universities and schools were created in a number of states. After 1945, an actual network of federal universities was established, covering almost all states. The first Catholic university, the Pontifícia Universidade Católica do Rio de Janeiro, was created in 1940. In the following years, new Catholic universities were founded in major state capitals all across the country. Separate nonuniversity institutions also continued to be created by state governments and through private initiative.

THE TURMOIL OF THE 1970S

In 1968 the federal government, then under military rule, enacted legislation seeking to reorganize the entire Brazilian higher education system. The reforms replaced the old chair system with the department model, proposed the adoption of full-time contracts for faculty, regulated graduate education, and made the transition at the undergraduate level from the conventional sequential courses to the credit system, similar to the U.S. model (for an overview of the 1968 reforms, see Klein, 1992; and Durhan, 1998). The reforms faced resistance from faculty at the most traditional professional schools and mistrust from many faculty and students, due to their authoritarian origins. Nevertheless, in the long run the reforms were successfully implemented in the entire public sector, at both the federal and provincial level. Estimates made by specialists show that the universities' share of the federal budget grew 5.4-fold between 1972 and 1986, most of the increase consumed by the implementation of full-time contracts among faculty (Schwartzman, J., 1993; and Velloso, 1987).

The 1968 reforms were implemented amid an explosive increase in the demand for higher education. In 1960, the total enrollments in Brazilian higher education stood at 93,000 students (Schwartzman, S., 1992). In 1970, the enrollments had already jumped to 425,478 students. Five years later, in 1975, students at the undergraduate level numbered 1.1 million. This greatly enlarged student population was not foreseen at the time of the 1968 reforms. In order to respond to this demand, the government relaxed the constraints over the creation of new institutions, especially nonuniversity ones. The new colleges and professional schools then created through private initiatives absorbed the bulk of the expansion, preserving the public sector from the most deleterious effects of mass higher education. The growth of the private sector was achieved mostly by an increase in the number of for-profit, teaching-oriented, nonuniversity schools and colleges. Low-quality faculty, paid on an hourly basis, staffed them.

In the public sector, entrance examinations were (and still are) used to control the growth of enrollments and thus limit the pressures with regard to teaching load. Full-time, stable contracts with low teaching loads were almost universal, even in those universities unable to secure a high academic profile for their faculty. Graduate education and research in this sector experienced an even more dramatic increase after the 1968 reforms. Enrollments at the graduate level rose from almost zero in the early 1960s to almost 100,000 students by the end of 1990s.

In the late 1960s, the federal government directed huge resources toward the graduate level that was being created at the most prestigious universities. Money was provided to support programs, and fellowships were offered to attract students. Fellowship programs were also instituted to raise the qualifications of professors through graduate studies domestically and abroad. And, contrary to the experience at the undergraduate level, the government and the academic community made a decisive effort to assure quality at the graduate level. The Fundação Coordenação de Aperfeiçoamento de Pessoal de Nível Superior (CAPES), the Ministry of Education agency in charge of graduate education, created a sophisticated peer review evaluation system that connects performance with support at this level.

Direct support for high-ranking graduate programs—provided by CAPES and other agencies, bypassing the university bureaucracy—allowed them to engage faculty with Ph.D.s earned abroad. Such programs became the main sites for the institutionalization of research in the Brazilian higher education system (Oliveira, 1984). Nevertheless, the very success of these programs was also a source of their weakness. Between 1970 and 1980 the support for graduate education was so great that many graduate programs effectively separated from the departments, leaving undergraduate education without the input of the most qualified faculty (Castro, 1985).

The Profile of the Higher Education Enterprise

Stratification

All these developments have made Brazilian higher education not only a diversified system but also a highly stratified one. At the top there are a small number of universities, most of them public, strongly motivated by academic standards, where research is a permanent and fully institutionalized activity. In the first tier, institutions provide a better working environment, which in turn allows them to attract better-qualified academics, as well as financial support for research. These institutions award the majority of the doctoral degrees granted in the country.

The second tier includes most public institutions. Most of these are universities, but lack the minimum conditions for effective academic development. They have not been able to establish strong graduate programs and thus have problems attracting and retaining doctorate holders among their faculty. In this tier the bureaucracy and unions have greater power, and administrators also tend to have more room for taking initiatives.

At first- and second-tier universities, prestigious private foundations carry out the applied research, outreach, and consulting activities of the universities' professional schools. These foundations were created at the initiative of the faculty to bypass the bureaucratic constraints imposed by the country's public spending regulations. These foundations allow professional schools to offer income supplements to their faculty and are an important source of fund-raising. Foundations are formally independent of the university and usually have a formal agreement with the professional school, paying a fixed percentage of their total income as overhead to the university's central administration.

The third and lowest tier comprises the majority of Brazilian higher education institutions. They are mostly private institutions or owned by small municipalities and poorer states. Most of them are small colleges or separate professional schools. But there are also giants among them, holding the status of universities, where undergraduate enrollments can reach 30,000 students or more.

The 1968 reforms had little or no impact over the institutions in this tier, nor were they supposed to have one. Such institutions are not organized on a departmental basis but are vertically structured, with the smaller unit being the bachelor's program. Supervising each program is usually a coordinator, who is often a senior faculty member with long years at the institution. The coordinator's authority derives from the trust of the institution's owner[2] rather than from his or her academic reputation.

These institutions are very successful in meeting labor market demands for quick training. Some of the poorer ones are only motivated by the opportunity to provide credentials for specific jobs.[3] While others are committed to quality, the criterion used to measure this quality is the employability of their alumni (Sampaio, 2000). The permeability of the academy to the outside pressures of the job market has important and beneficial consequences for some institutions in this tier. They are more able to respond quickly to new demands from the labor market, and in some cases are very successful in offering training alternatives for working professionals and good guidance and placement opportunities for their alumni.

Lastly, an important point should be stressed: None of the tiers sketched above can be linked to just one sector of the Brazilian higher education system. As stated earlier, in the top tier one finds federal universities, state-owned universities, as well as some nationally well-reputed private universities, such as the Pontífícia Universidade Católica do Rio de Janeiro. In the second tier, one finds most of the federal and state-owned institutions,[4] but also some private, mostly Catholic institutions. The third tier consists mostly of private institutions, but also some public institutions owned by small municipalities and poorer states.

STRATIFICATION: SOME MEASUREMENTS

Table 4.1 shows the distribution of enrollments at the undergraduate and graduate levels among the different tiers and the proportion of Brazilian academic professionals employed in each tier.

Table 4.1 Brazilian higher education enrollments, by tier (percentages)

Tier	Undergraduate enrollments	Master's level	Doctoral level	Academic professionals
First tier	5.4	29.6	57.2	7.9
Second tier	27.8	56.7	36.5	39.7
Third tier	66.8	13.7	6.3	52.4
Total N	2,124,054	69,702	30,302	164,900

Note: From MEC/INEP, 1998 (tabulations made by authors).

Diversification

Few studies have been undertaken on the conditions of the Brazilian academic workplace. The most comprehensive study was carried out in 1992, with support of the Carnegie Foundation for the Advancement of Teaching (Schwartzman, S. and Balbachevsky, 1996). In this survey of the Brazilian professoriate, a total of 898 academics were interviewed in a random, stratified sample of Brazilian states, higher education institutions, and academics. The sample included professionals with full-time and part-time contracts, selected from the faculty lists provided by the institutions. The intention was to cover the diversity of sectors, types of institutions, and regions in the Brazilian higher education system. However, the study oversampled academics belonging to the first tier. Nevertheless, an analysis of the data produced by this sample does reveal some important differences in the academic workplace among the three tiers.

The first striking difference between tiers involves faculty opportunities for promotions in rank, as determined by academic degree. Table 4.2 presents the institutional ranks and academic degrees reported by faculty included in the Carnegie sample.

Table 4.2   Rank distribution, by academic degree and tier (percentages)

	Academic degree			
	Master's or below	Doctorate	*Livre docencia*	Total
First tier				
Assistant	90.7	72.3	—	66.7
Associate professor	6.7	23.4	68.6	25.0
Full professor	2.7	4.3	31.4	8.3
Total *N*	75	94	35	204
Second tier				
Assistant	53.9	7.7	—	41.9
Associate professor	41.6	80.8	65.5	50.6
Full-professor	4.5	11.5	34.5	7.5
Total *N*	401	104	29	534
Third tier				
Assistant*	56.3	42.9	—	52.4
Associate professor	15.8	—	7.7	14.8
Full-professor	27.9	57.1	92.3	32.9
Total *N*	190	7	13	210

Note: From Schwartzman, S. and Balbachevsky, 1996; and Balbachevsky, 2000. The dashes indicate that no cases were found.
* This line includes academics paid on an hourly basis.

To understand the meaning of the data in this table, it is important to understand how the Brazilian graduate level is organized. This level encompasses two stages: the master's program and the doctoral program. Master's programs are considered to be an intermediate stage in the training for academic life and not, as in the Anglo-Saxon context, the terminal stage for professional life. This pattern also applies to professional fields like law and engineering. The doctoral degree is also an academically oriented degree, without special value outside academe. In the latter half of the 1990s the master's degree acquired a value on the job market outside academe. However, this is a recent trend in the Brazilian job market, not officially recognized by CAPES or accepted by the academic community. Since 1997, CAPES has been trying to implement a new kind of master's program—the professional master's program—one oriented toward the demands of the job market with regard to graduate training. In the CAPES proposal, this program should include professionals outside academe as faculty and should be self-sustaining—meaning that the funding from federal agencies would be supplanted by revenues from the tuition it would be allowed to charge. Among academics, on the other hand, the implementation of such programs has evoked considerable resistance. The main argument against such programs is that they would open up the public system to market pressures.

Master's programs require students to complete a variety of courses and work on a short dissertation under the supervision of a qualified adviser. As in the French model, the dissertation is supposed to be defended in a public session, with two peer examiners (Neave, 1993). The doctoral program is considered the final stage of training for researchers. Candidates at this level of training are expected to have successfully completed the master's level. Doctoral training takes from four to six years, while the master's program takes from two to four years (Velloso and Velho, 2001). The doctorate also involves completing a sequence of courses and producing a major dissertation. Once again, the final step is the public defense of the doctoral dissertation, before a committee of five experts: the adviser and four other professionals, with at least two from outside the program.

The *livre docencia,* which is an adaptation of the German academic title of *Privatdozent,* is a degree attained after a public examination and the presentation of an academic dissertation. However, this degree is not linked to graduate programs and is not evaluated by CAPES. It is an institutional title recognized by many Brazilian universities. At some first-tier institutions candidates for the *livre docencia* are also required to hold a doctoral degree, although this is not the case at other institutions. At many institutions the *livre docencia* is a mechanism to circumvent the doctoral degree legal requirement for promotion (Schwartzman, S., 1992).

As one can see in table 4.2, at first-tier institutions academics without doctoral degrees tend to remain in the lowest rank. Here, 91 percent of faculty without a doctoral degree are employed as assistants. At some institutions, even the doctorate is no guarantee of promotion. At these institutions, the *livre docencia* remains a formal postdoctoral degree, required for promotion to full professor and, in some cases, even to the post of associate professor. In fact, 72 percent of academics with a doctorate reported their rank as assistant.

At second-tier institutions, requirements for promotion are not as strict: A fair proportion of faculty with a master's degree or less (46 percent) are promoted to the position of associate professor or higher positions. On the other hand, holding a doctorate is a sufficient precondition for promotion at institutions in this tier. Among professors with doctoral degrees, 81 percent reported their rank as associate professor and 11 percent as full professor. In the third tier, the links between degrees and ranks are blurred. Promotion reflects mostly seniority and the institution's trust in the professional.

Another important dimension in the institutional environment relates to job stability. The Brazilian higher education system does not have tenure as it is known in the United States and the United Kingdom. Nevertheless, at public institutions, faculty are supposed to hold stable contracts as a result of their status as civil servants. In the private sector, job stability is seldom recognized, although some academics do regard themselves as holding stable contracts due to special arrangements they have with their employers. Also, the most prestigious Catholic institutions recognize job stability as a benefit attached to the higher positions in the institution's career path. Table 4.3 shows how Brazilian academics perceive the condition of stability regarding their contracts.

Although stability is a benefit associated with civil service status, institutions in the first tier, even public ones, do link this benefit with academic performance, as measured by the academic degree. As one can see in table 4.3, only 34 percent of faculty with a master's degree or a lower degree in first-tier institutions report having stable contracts. For academics with doctorates, the percentage is significantly higher, 61 percent, while among the *livre docentes* 89 percent had achieved job stability. The recruitment process at these institutions usually requires a probationary period, during which the professional

Table 4.3 Job stability, by academic degree and tier (percentages)

	Academic degree			
	Master's or below	Doctorate	*Livre docencia*	Total
First tier				
Stable contracts	34.2	60.6	88.9	55.8
Nonstable contracts	65.8	39.4	11.1	44.2
Total *N*	76	94	36	206
Second tier				
Stable contracts	84.7	84.6	93.1	85.1
Nonstable contracts	15.3	15.4	6.9	14.9
Total *N*	398	104	29	531
Third tier				
Stable contracts	18.6	—	25.0	18.4
Nonstable contracts*	81.4	100	75.0	81.6
Total *N*	147	7	12	207

Note: From Schwartzman, S. and Balbachevsky, 1996; and Balbachevsky, 2000. Dashes indicate that no cases were found.
* This line includes academics paid on an hourly basis.

is connected with the institution by temporary contract. These temporary contracts entitle academics to all the privileges and duties of the civil service—including full-time employment, the same salary levels, and the standard teaching loads—but without stable contracts. This pattern of recruitment is not enforced by the institution's central administration but represents a practice chosen by most institutes and departments in the universities in this tier.

At institutions in the second tier, job stability is a widespread benefit and is not connected with an academic degree. In this tier, 85 percent of faculty reported having stable contracts, regardless of their academic degrees. Obviously, this high percentage of faculty on stable contracts reflects the predominance of public (federal or state-owned) institutions in this tier. But what is relevant here is that in second-tier institutions, stable contracts are regarded as a right to be assigned to all faculty, regardless of academic achievement or experience.

In the third tier, job stability also has no relation to academic degree. However, here, nonstable contracts are the norm: Only 18 percent of academics employed in this tier have stable contracts. This figure reflects the high proportion of private institutions in this tier.

Another aspect of academic employment in Brazil is the high percentage of faculty with full-time appointments at public institutions (see table 4.4). In 1992, in the first and second tier nearly 70 percent of the professionals reported having full-time contracts. Throughout the 1990s this figure increased. Even in the third tier, where full-time faculty were rare in 1992, the proportion of professionals with full-time contracts increased quickly over the decade. In the public sector, whether one has a full-time contract is a matter of personal choice and also depends on the options available in the respective academic field. It is the preferred path for academics working in the basic sciences and the humanities. But the figures in the technological and social sciences and professions are significantly lower. Furthermore, in the first and second tiers, the proportion of faculty with full-time contracts shows little variation by rank or degree (Balbachevsky, 1999).

In the public sector, full-time contracts ensure a good income level for faculty. It is estimated that a professor with a doctorate, employed in the federal sector, earns an income similar to that of a middle-rank manager in industry. Salaries in the public sector, even at universities owned by the states, tend to follow federal standards. Usually, a doctorate holder in the public sector earns approximately U.S.$20,000 a year. But wide differences exist, based on seniority and grievances won by unions at some institutions. Differences in payment between ranks are small, so even without a doctoral degree a professor in the

Table 4.4 Academics reporting full-time contracts

Institution	Percent
First tier	68.6
Second tier	70.2
Third tier	2.3

Academic field	Percent
Physical sciences	83.7
Biological sciences	73.3
Social sciences	52.9
Humanities	58.7
Engineering	33.6
Health professions	46.3
Administrative and legal professions	38.6
Technical professions	40.6
Education	54.3

Source: From Schwartzman, S. and Balbachevsky, 1996.

public sector has good prospects with regard to standard of living. Thus, professors in the public sector are rarely found holding other academic positions.

One must be careful when comparing incomes in dollar figures. While the cost of living calculated in the local currency has been stable since 1994, rising exchange rates have been a factor in the Brazilian economy over the last two years. Therefore, an annual income of approximately U.S.$20,000 allows for a comfortable middle-class standard of living, especially in a two-income family, which is the norm in the Brazilian middle class.

Nevertheless, income is a significant source of dissatisfaction among academics in Brazil. Complaints about the decline in academic salaries are often heard at public universities. Since the implementation of the 1994 stabilization plan, important horizontal increases in academic salaries in the public sector are more or less out of the question because of federal policies emphasizing fiscal equilibrium. On the other hand, academic salary levels in the public sector are rigidly regulated. Institutions are bound by external regulations and can do little to reward deserving faculty.

The prestigious private foundations linked to the professional schools at the best public universities usually provide salary supplements, which helps to make the academic market in the public sector attractive even in fields that are well paid in the general labor market. Besides, in the more market-oriented fields, being a professor at a prestigious public university creates opportunities for consulting and community service, which in turn engender new sources of income for these professionals. But these ways of supplementing academic salaries lack legitimacy inside the university. For example, the unions have always opposed them as a tool for introducing differentiation among faculty, which is an anathema for them.

Faculty salaries for full-time contracts in the private sector are usually higher than in the public sector. But at private institutions, full-time contracts are fewer in number and include strict requirements imposed by the institution. Professors with such contracts usually carry a heavy teaching load, up to 20 to 25 hours per week, as well as considerable administrative duties.

Table 4.5 shows the allocation of time in a typical week, as reported by the faculty members interviewed in the Carnegie survey. At most Brazilian public institutions there are active groups of professors who work intensively in research and teaching. On the whole, however, only at institutions of the first tier does one find a similar balance in the reported allocation of time spent on teaching and on research (on average, 11.5 hours per week). This finding holds true even when rank and degree are taken into account (Balbachevsky, 1999). At institutions in the second tier, teaching is the more time-consuming activity (on average, 15 hours per week), while time dedicated to research falls behind (7 hours). In this tier, the academic degree introduces an important distinction: Faculty with master's or lesser degrees tend to spend more time teaching than doing research, while professors with doctorates tend to dedicate an equal amount of time to research and teaching (Balbachevsky, 1999).

Regardless of the great amount of time professors in the first and second tiers spend on teaching, the teaching load is relatively low in the public sector. The Brazilian education act sets an eight-hour-per-week teaching load for faculty with full-time contracts at public institutions, regardless of academic degree and rank. In the most academically oriented universities, teaching loads are even lighter due to the importance of the graduate level. Furthermore, beyond teaching load, an institution's central authorities have little control over professors' schedules. It is mostly the internal structures, such as departments or laboratories, that regulate the way professors spend their time. Thus, one finds huge differences among departments at the same university concerning the intensity of the professors' commitment to research, publishing, consulting, and so on.

In the third tier, less time tends to be spent on teaching than on activities outside the academic world. On average, academics employed in this tier reported spending 13 hours on teaching and 14 hours on other, nonacademic, activities. Research occupies significantly fewer hours (3 hours) in this tier.

Table 4.5 Hours spent per week on professional activities (average numbers)

	First tier	Second tier	Third tier
Teaching	11.5	14.8	12.8
Research	11.4	7.3	3.4
Service	3.8	3.6	5.3
Administration	3.3	3.3	1.3
Other academic activities	1.8	1.6	0.8
Nonacademic activities	3.2	4.3	14.4

Note: From Schwartzman S. and Balbachevsky, 1996; adapted from Balbachevsky, 2000. Answers were adjusted to a standard 40-hour work week. When the total of hours reported was less than 40, the difference was attributed to "nonacademic activities." When it was more, values were proportionally reduced to fit the total of 40 hours.

The figures in table 4.5 reveal a picture of the highly diversified niches one can find in the Brazilian academic workplace. At the top-tier institutions, full-time contracts are the rule. The other special trait of the institutions at the top is the high proportion of doctorate holders among their faculty. Among the institutions in the top tier, institutional ranks correspond strictly to the academic degree (Balbachevsky, 1999). Apart from the lowest position of teaching assistant, there is nowhere to go without a doctoral degree. The institutional ethos strongly values research activities and participation in the graduate level, this is the case even in the less competitive departments.

In the top tier, departmental life is dynamic and faculty collegial power strong. This allows for very diversified niches of working conditions. In the most competitive environments, especially in the hard sciences, departments are structured around laboratories headed by full professors with strong academic and scientific reputations. Resources to sustain research flow toward laboratories and graduate programs, on the initiative of researchers working mostly in highly structured teams.[5] In the less competitive departments, the low-keyed academic environment allows faculty to move more freely. In such environments research tends to be an individual endeavor, with little departmental support. As research teams form, they usually manage to free themselves from departmental authority. The department's ability and will to support research activities are deemed unstable, due to the internal divisions and a blurred pattern of authority.

At institutions in the second tier, full-time contracts are also widespread. As in the case of first-tier institutions, part-time contracts are a personal option, not a restriction imposed by the institution. Here, however, professors with doctoral degrees are a scarce resource. In most of these institutions, such professors represent less than 20 percent of the faculty. In such institutions, academics with better credentials tend to be concentrated in certain departments, creating unusually dynamic microenvironments within the institution. They truly represent islands of academic competence and are responsible for the graduate programs organized in the institutions.

In this tier, the reward structure generally emphasizes seniority rather than academic performance. Being formally recognized as public universities creates high expectations among their faculty concerning status and working conditions. Even without high academic qualifications they enjoy access to an academic career, full-time contracts, and low teaching loads. It is among the faculty at these institutions that one finds the greatest resistance to any reform. For them, any change designed to move the system into a more competitive environment is a threat.

Finally, there are the institutions in the third tier, in which full-time contracts and professors with doctorates are seldom found. These institutions are staffed by teachers paid on an hourly basis. Some staff pursue teaching as a complementary, often secondary, professional activity. This group includes lawyers, economists, engineers, architects, physicists, and others, whose nonacademic professional stature is an important source of institutional prestige. On the other hand, being a higher education faculty member gives them social distinction and enhances their professional reputation. Other faculty at third-tier institutions have teaching as their principal professional activity. They earn their living by teaching a great number of classes, with heavy individual teaching loads that can reach up to 30 hours per week.

The large institutions in this tier also have impressive administrative bodies, whose administrative staff have an important voice on the future of the institution. Faculty and union perspectives find little expression. Rank is based largely on institutional trust rather than on academic or seniority considerations. Being more centralized, third-tier institutions often have greater flexibility and scope to respond quickly to the new pressures put out by the environment.

CHANGES IN PUBLIC CONCERNS

As stated in the opening of this chapter, the 1990s were a decade of major economic reforms in Brazil. Changes in the framework of the economy had a powerful impact on the Brazilian higher education system's demand side. Table 4.6 outlines the shifts in values, attitudes, and concerns present in Brazilian society regarding education policies, from the 1970s through the 1990s.

On the basis of this table, one might describe the changes in attitudes toward Brazilian higher education policy as moving in two directions: away from an elite-formation perspective and toward a general workforce-qualification perspective. Thus, we see a combination of worries about the quality of undergraduate programs, the employability of alumni, general science and mathematics literacy, and the quality of teacher education. There is a trend toward increased demand for quality control. As one can see in table 4.6, in the 1970s this concern was perceived only at the graduate level. In the 1990s, this concern focused on the undergraduate level, including the mass-oriented (i.e., relatively nonselective) private sector, and teacher training programs.

Table 4.6 National concerns about higher education

Concerns	1970s	1990s
University institutional autonomy	low	medium
Control of public spending on higher education	low	high
Quality of undergraduate education in the public sector	low	high
Quality of undergraduate education in the private sector	low	high
Quality of graduate courses	high	high
Diversity of science and technology human resources	high	high
General science and mathematics literacy	low	medium
Outputs of scientists	medium	high
Elite formation and enlightenment	high	low
University-productive sector interface	low	high
Teacher education	medium	high
Employability of alumni	low	medium
Faculty and alumni entrepreneurship	low	high
Regional equity	high	high

Note: Based on the author's analysis of the literature and official documents; adapted from Diederen et al., 2000.

The other direction public concerns have taken is the shift from an orientation toward inherent academic values to a more demand-driven one. Expectations are increasingly focused on the output of scientists, the interface between the university and the productive sector and the beneficial impact of knowledge on the competitive edge of businesses in Brazilian society.

GOVERNMENT RESPONSE

In the past, government response to public concerns has been to increase controls over the institutions. In the new environment, the government is being called upon to answer for the quality of the training offered by all institutions in the system.

In 1994, the Ministry of Education took its first steps toward implementing an effective evaluation of undergraduate courses. In 1995, the ministry introduced the National Undergraduate Programs Evaluation, consisting of a mandatory national examination that measures the performance of all students in the same career path. Even though individual student scores are not made public, the average performance of students in each institution is widely publicized and has a great impact on public opinion.

The outcome of the national exam is used by the Ministry of Education to rank institutions in each career path. The ranking procedures take into consideration the institution's average student performance, its infrastructure, and faculty academic profile. The institutions that place in the rank's bottom quartile of the rankings are put under supervision. An Expert Visiting Committee suggests corrective measures. Failure to achieve the recommended improvements is punishable by withholding the undergraduate program's official permit.

The total impact of this evaluation procedure has been impressive. Poor performance on the exam has forced institutions in all sectors and tiers to begin focusing on quality at the undergraduate level. The effect of the national exam has been more visible in the private sector, which is subject to the strict control of the federal government.

Administrators in the private sector are very aware of the federal initiatives, although their responses have been diverse in nature. Brazil has an entrenched legacy of evasive, bureaucratic forms of evaluation. In practice, the government demands indicators, and institutions simply invent the figures that make them look good—a stratagem most private institutions have continued to employ. They try to achieve better results on the national exam by coaching students, to raise the average academic qualifications of faculty by hiring a few doctorate holders, and to improve infrastructure evaluations by constructing a few ostentatious buildings. These investments are intended to satisfy the demands of the ministry's Expert Visiting Committees but have little bearing on the institution's teaching needs.

Yet a number of institutions have taken up the challenge to improve their quality of teaching. They have prepared comprehensive plans to enhance institutional quality (including total quality in education), put in place ways of rewarding faculty when student scores on the national exam improve, and begun to attract qualified professors by offering support for their research or consulting activities. While it is too early to assess the net results of such efforts, one can see that the private sector is introducing changes.

In 1997 the Brazilian government also enacted a new education act, the Lei de Diretrizes e Bases da Educação (LDB). The LDB, while declaring that research, teaching, and community service are the primary goals of higher education as a system, explicitly recognized—for the first time in recent Brazilian history—the existence of institutions primarily devoted to undergraduate teaching. In the past, the ideal of a unitary system, governed by the same rules and sharing common goals, informed all Brazilian higher education regulations. Recognizing differences is important, since it frees institutions from the weight of bureaucratic regulations that only waste energy.

The new education act also granted expanded autonomy to universities while increasing the focus on their academic profile. According to the LDB, in order to be accredited (and, for the first time, reaccredited every five years) as a university, at least one-third of an institution's faculty must have a master's or higher degree, and academic titles must form one of the necessary criteria for career promotion.

For the federal universities, the government proposed, in effect, financial autonomy. In the government proposal, budget increases for each university would be connected to increases in performance. This proposal was received with mistrust by university authorities, faculty, and employees and still lacks support among politicians. Some stakeholders were concerned that the proposal failed to obtain the necessary commitment of officials in charge of Brazil's macroeconomic policies. Others suspect that university autonomy could be the first step in a government plan to privatize the federal universities. Still others fear autonomy because it could give rise to competition within and among institutions. Distrust of the government's intentions is coupled with regional concerns that autonomy could mean the transfer of the financial burden for operating the universities from the federal sphere to the states. So, as of 2002, the convergence of these diverse forces succeeded in stymieing any move on the autonomy project.

The government has implemented some other measures at the federal universities, establishing an extensive fellowship program that rewards professors for their commitment to undergraduate programs. Fellowships in this program are given to the institutions, which are responsible for awarding them to faculty. This process was intended to introduce a modicum of competition into the institutional environment. However, most institutions have succeeded in avoiding this last effect, finding ways to circumvent the conditions put in place by the government.

At the graduate level, CAPES is implementing more stringent evaluation procedures. At this level, the most important goals are, first, to reduce the time required to complete both master's and doctoral programs; and second, to improve the quality of the top programs, submitting them to international peer review.

In spite of the obstacles, these initiatives to change the system's regulatory procedures have achieved some important results. The effort has promoted differentiation and moved the system as a whole to raise its standards. Improvements are more noticeable in institutions situated in the third tier. In the private sector, a number of institutions have succeeded in their attempts to be certified as universities, upgrading the academic profile of their faculty, and creating an institutional career path where none existed.

A few institutions in the private sector are also trying to become active as research centers. They usually hire retired professors from the public sector to supervise research units or direct laboratories. These institutions prefer to work in interdisciplinary fields where the competition is less intense. Until now, the success stories have been few in number, but those few are very impressive.[6]

Institutions in the first and second tier have improved their faculty academic profiles. Pressures brought to bear by the federal regulatory agencies have resulted in more rigorous institutional recruiting policies, raising the threshold for academic qualifications. All in all, these new developments tend to create a more competitive environment for higher education as a whole (Sampaio, 2000). Nevertheless, in the long run, the net outcome of these changes is difficult to predict. Much depends on the ability of higher education institutions to respond to the challenges presented by the new environment. At the institutional level, conflicting pressures create unstable surroundings, which is the source of the discomfort expressed by academics in the inquiries made for this study.

Challenges at the Institutional Level

The first problem faced by most public institutions involves institutional governance. Brazilian public universities have traditionally adopted the European model.[7] The highest authority within the university is a collegial body—the university council (*Conselho Universitário*), elected by faculty, students, and employees. The administration is under the control of the university's rector. It is the rector's responsibility to make appointments to top administrative positions, selecting persons from among the faculty.

The democratizing process[8] that took place in Brazilian society from the end of 1970s to the end of 1980s forced the incorporation of demands from students, assistants, and employee unions for a one-man-one-vote electoral system for appointing the rector. Almost all public institutions adopted such a system in one way or another. The democratizing process also entailed the idea that universities should be free to select their administrators internally, without the help of any kind of external body or consulting process. After more than two decades of such "democratic" governance, some problems have been widely acknowledged (Schwartzman, S., 1992; Coelho, 1988; and Castro, 1985).

In the absence of an external source for their mandates, rectors became trapped by the conservative forces that usually oppose any reform threatening the status quo. Yet, the situation—often combined with limited financial autonomy—produced few incentives for increasing the accountability of the university's management. Finally, the populist environment created by such practices is not conducive to the needs of a sound academic reward structure able to recognize performance and support differentiation among academics. The reforms of the 1990s did not address this problem.

Another important challenge is the weight of the bureaucratized structures and management procedures at the public universities. Since the early 1990s, Brazilian public universities have been pushed externally toward role diversification and accountability. Brazilian society has called for changes in public-sector higher education with regard to teaching standards, employability of alumni, continuing education programs, public precollegiate education, and the interface between the university and the productive sector. The quality of the public sector's response to these demands for reform will determine how this sector is evaluated in public opinion and by government authorities.

At the same time, public universities are themselves poorly equipped to answer these demands. A tight departmental structure hinders interdisciplinary efforts. An overly internal orientation and structural rigidity make it difficult for this sector to recognize and reward faculty initiatives that do not fit within the traditional academic parameters. The usual way in which universities deal with external pressures is to tighten bureaucratic controls and centralization. This is a poor pattern of behavior, when initiative, flexibility and innovation are required. So, in our interviews, not surprisingly, we found dispirited academics working in an environment under challenge from the new needs and demands of society.

In the top tier, public institutions have faced an additional challenge: the internationalization of their scientific endeavors. In 1990, a study of the Brazilian academic profession found that even among first-tier institutions, only 21.6 percent of faculty had some research interaction with foreign peers over a three-year period (Balbachevsky, 1999). This pattern of relative detachment from global academic research continues. Even the impressive graduate-level programs established since the 1970s remain relatively isolated from their counterparts abroad. In recent years some improvement has occurred, but until now this has been centered in a few hard-science fields, such as genome research. In most other fields, progress in this dimension has not kept pace with the globalizing of science and technology.

Another challenge concerns the internal career path and reward structure of higher education. In the private sector, career path has until now had little meaning. As a rule, part-time teachers staff these institutions. The career stages, which do exist, reflect more the institution's reliance on an individual teacher than on the person's academic or professional performance. While this picture has started to shift in the last few years, most private institutions are poorly equipped to deal with the new situation. Because of their financial constraints, they cannot copy the expensive public-sector solutions, and their lack of knowledge about international experiences leaves them with few clues as to how to manage this issue.

At public universities, the institutional career path is linked to academic degrees, but to little more. The career path includes only a few stages and is regulated by strict bureaucratic procedures. Even at the top universities, evaluation procedures do a poor job of considering indicators other than publications and involvement with the university's bureaucratic structures. In the least competitive departments, a person's career is only loosely linked to performance. In the most competitive ones, internal pressures make a faculty member's career more dependent upon actual academic productivity.

At public institutions of the second tier, requirements are even less strict. The internal environment provides numerous avenues to bypass the official requirement of academic degrees. These universities lack the incentives to attract more research-oriented academics. On the other hand, they have little room to diversify and search for alternative factors in faculty evaluation and institutional pride, being a part of a sector strictly regulated by uniform norms.

The last institutional challenge, faced by all sectors, concerns funding. The 1990s have been marked by the success of the government stabilization program—the Plano Real. But the impacts of stabilization were, in the short run, not beneficial to the Brazilian higher education system. In the federal system, the lack of financial and administrative autonomy, coupled by government pressures for stricter budget control, has had harmful effects—especially concerning the ability to maintain and improve infrastructure. Money for these needed improvements has been scarce since the early 1980s. Nevertheless, during periods of high inflation, the federal universities financed these investments by submitting to the federal treasury an inflated budget for salaries at the beginning of the year. Since the surplus remaining after actually paying the salaries was to be returned to the treasury at its nominal value (without monetary correction) and not until the end of the year, this generated a surplus in real terms. It was this difference that allowed university officials to make investments in and to maintain the infrastructure.

Given this precarious solution, it is not difficult to understand the overall negative effects of monetary stabilization on the federal system. For this sector, stabilization meant impoverishment. Since 1998, the Ministry of Education has sought ways to alleviate the situation. But controls over the education budget are tight, and new issues—such as basic education—compete with higher education on the ministry's policy agenda.

In the private sector, stabilization created strong pressures for more effective budget controls. In times of high inflation it was relatively easy to transfer increases in costs to the tuition charged. Money could also be generated almost magically through maneuvers in the overnight financial market. This state of affairs underwent a radical change after stabilization. For some institutions, especially the smaller ones, past mismanagement, bad investments, and a high incidence of breached tuition contracts have taken their toll, and the very survival of these institutions is in doubt. Other institutions fared better in the new environment. They diversified their portfolios of courses, exploring new market niches such as continuing education programs, nonacademically oriented graduate education, training programs tailored to meet the needs of special clients in the productive sector, and so on.

CHANGES IN THE ACADEMIC WORKPLACE

The most striking change in the Brazilian academic workplace relates to the shift in the relative market share between the public and private sectors, especially in regard to professionals with graduate degrees. Until the 1990s, most of the private sector did not hire qualified academics. For institutions in this sector, academics with degrees were an unaffordable luxury, and in any case they had no use for the title professor. What mattered was their students' employability rather than the qualifications of their faculty.

After the 1995 education act, the Ministry of Education focused increased attention on the academic credentials of faculty. The act established that at least one-third of the faculty must have the master's degree for an institution to be accredited as a university. Accreditation as a university lends prestige and permits the institution to function with greater autonomy. Only universities have the authorization to start new undergraduate programs without asking the permission of the ministry. In addition, the ministry's official ranking of undergraduate programs includes an assessment of the faculty members' academic qualifications. Having a good share of master's and doctoral degrees among the faculty is the best way to ensure a good position in the ranking.

All these initiatives combined to create a new market for young, titled academics. In the mid-1990s, to gain control over the public spending that was regarded as a major source of inflationary pressures, the federal government forbade the hiring of new civil servants. With the lack of new faculty places at public universities, the private sector presented an attractive alternative to academics. But hiring this new kind of professional created new challenges for the institutions in the private sector and entailed a differentiation among faculty and pressures for research support and rewards in terms of individual career paths within the institution. With the new regulatory framework in place, the total number of full-time contracts in the private sector increased, although the percentage of academics with full-time contracts in the sector remains low, usually between 5 and 20 percent of all faculty. At the larger private universities the proportion of full-time contracts is higher, from 15 to 30 percent of all teachers. But even 5 percent would have been unheard of in the past, when almost no faculty at private institutions had full-time contracts.

In the public sector, first- or second-tier institutions, the changes outlined above have had little impact. The Brazilian public universities, which are well established, have until now been successful in resisting the most important pressures put upon them. Even when top administrators may wish to respond to the changing environment, they have little room to take strong measures. For their part, faculty bodies are typically slow to act.

In fact, the most important changes in this sector have been produced by forces and movements outside the system. Economic stabilization has had a major effect on the income levels of members of the middle class in Brazil, for whom high inflation rates had been a source of profit in the past—with their incomes indexed and their money in the overnight financial market. The first years of economic stabilization were hard ones that affected academic professionals, as part of the middle class. Professors employed in the public sector also witnessed a freezing of their salaries, due the government's reluctance to allow any horizontal increase in the salaries of the civil service. Nevertheless, the stable currency made the absence of income increases tolerable for academics.

Also between 1995 and 1998, while debates were under way on reforms of the social security laws, especially those concerning retirement, many professors in the public sector started retiring early—most of them between 40 and 50 years of age—in order to avoid the new regulations.[9] As some of the retirees were prominent professors and leaders in their departments, this trend introduced critical changes in some places: Some leading centers of research were disrupted by the early retirement of key figures, and well-known graduate programs were weakened by the departure of prominent professors and experienced advisers. Nevertheless, in the long run, the outcome was not so bad. Even though some good professors retired, the majority of these early retirees were older professors without strong academic credentials. In 1998, when public universities started hiring faculty again, the new places were usually occupied by young academics, most with a doctoral degree, who brought fresh academic perspectives to the institutions.

On the other hand, some of the experienced professors and researchers who retired from the public sector were hired by the private sector. They were attracted by good income prospects and the promise of support for new research initiatives. This trend helped to improve the academic profile of the institutions and introduced a new dynamism into the private sector.

In spite of the relative calmness with which the changes in the system have been absorbed, academics have felt the strain of coping in the new environment. The lack of a compass to guide them between the institutional environment and the new challenges put out by society is a source of uneasiness to academics, especially those working in the public sector.

FACULTY PERCEPTIONS AND CONCERNS

To explore how academics perceive the crosscurrents in the new environment, open-ended interviews were undertaken in late 2000 (from one to three hours in length) with a representative sample of 12 academics.

As the contents of the interviews were analyzed, it became clear that one of the factors determining how an academic perceives the changes in the institutional setting is the level of interaction between his or her undergraduate program and the labor market. The Brazilian higher education system is, even at the top institutions, a system oriented toward undergraduate education. Some undergraduate programs are structured as professional schools. They are externally oriented, and the professional ethos is more relevant than the academic ethos. Even when a professor has a full-time contract with the academic institution, she or he is essentially a professional in the market.

I'm an engineer. In engineering, isolation means death. The important point is having contact with the market. You must do consulting. With what you learn in consulting, you organize your teaching, be it in undergraduate or graduate programs. What would you teach your students if you didn't know what was going on in the market? (female, engineering, doctorate, top tier)

I'm in the field of business. What is business without the market? (male, business, master's degree, third tier)

The above comments show that for these academics, the undergraduate program, and not departmental affiliation, has become the principal affiliation the source of academic identity. Teaching, at the undergraduate level, is one of the most relevant sources of prestige in these fields, even at top-tier universities, where research and publishing are the most highly valued factors for institutional evaluation. At the top institutions, academics take pride in forming the profession's elite. In the third tier, the mobilizing quest is the future performance of students in the labor market.

The university asks for research, but our students (at the undergraduate level) are first-rate material. They are the best. We cannot risk losing this material. So, here the quality of the undergraduate courses is, and always has been, the top priority. From here, they (the students) will go on to the best posts.... Their performance, once in the market, will count for us. (female, engineering, doctorate, top tier)

At the top- and second-tier institutions, academics in the same undergraduate program have set up the so-called foundations to mediate the links with the market. As mentioned earlier, the foundations are independent, private institutions, identified with the school to which faculty are affiliated. Consulting, outreach activities—sometimes with nongovernmental organizations—training, and continuing education programs are conducted through these foundations. The link with the university is an important benefit for the foundation as well.

The foundation complements academic work. By working at the foundation, you can do relevant research, write papers everyone can read [even outside academic circles]. But it is the school that lends its name [and prestige] to the foundation. (male, economics, doctorate, second tier)

Holding a top executive position at a foundation is an important source of prestige and authority in these academic niches.

Here, the high prestige, which everyone is seeking and which results in power, is being at the foundation, doing consulting, and working with firms. Everyone wants to do research for firms, organize training programs with firms. It's not only for the money. Working at the foundation lends prestige. This is more important than just doing academic work. (female, business, master's degree, second tier)

However, academic credentials alone do not determine who succeeds at these positions. Success depends also on knowing the market and on building networks within it: "Knowing how to sell the knowledge you possess, who would need your skills, and how to reach those persons" (female, engineering, doctorate, top tier university). There is a hidden tension between values and working conditions in these academic niches and the values held by the institution as a whole. Complaints often focus on the evaluation process and career structure.

When your contract arrives there [at the internal commission for evaluating academic performance] the problem starts.... Everything I ask for is denied. They say I have a poor academic profile. Our work is poorly evaluated. They want publications, research projects supported by official agencies with peer review. Our clients support our projects (not the government)....When our graduate program received a "D" grade from CAPES, everyone here said: "It doesn't matter...." (female, engineering, doctorate, top tier)

Another source of tension concerns the fees and payments collected by the foundation and the "tuition-free" nature of Brazilian public universities. In Brazil, public universities are forbidden by law to receive tuition. As private institutions, foundations are free to collect fees and payments for their services and tuition for their training programs. This money helps to improve the school's infrastructure, supplements academic salaries, and so on.

We have good equipment here, space for conferences, multimedia and computing equipment—all paid for by the foundation. We didn't have to depend on the university money. The foundation also supplements academic salaries. So it helps us to retain good professionals who otherwise would leave the university. (male, economics, doctorate, second tier)

Not surprisingly, access to such resources provokes envy among the faculty outside the professional schools. But a more serious problem is that the foundations are in a dubious situation concerning their internal legitimacy.

> It's a problem, you know—everyone outside [the institute to which the foundation is linked] envies us. The faculty union wants to close the foundation down or forbid us from charging for our consulting. (male, economics, full professor, top tier)

At the other extreme, in terms of their perceptions of the changing environment, are the faculty working in the basic sciences and the humanities. In these fields, the sources of identity and pride are mixed. In the top and second tiers, most commonly the sources of both are found in the department, which provides the primary identification, especially when the academic is involved in the mainstream of the field in question. In the case of interdisciplinary fields, an academic's institutional affiliation is more ambiguous and complex. They may be considered outsiders in their departments. Thus, in interdisciplinary fields research centers or groups may become the primary options for professional identity. These nondepartmental structures are important places for graduate-level research training and are the only collective research environments open to graduate students working in interdisciplinary fields.

In hard science departments, laboratories and other structures for collective research are housed inside the departments. In the soft sciences and the humanities, such structures are usually located outside the department, which is viewed as an unsure setting for collective work:

> You know, some people do research. Others don't. If you bring in research money and equipment, you can never be sure it will remain with your team in the end.... There is always an uncertainty about the decisions taken by the departmental council. (male, political science, full professor, top tier)

Science and technology support agencies in Brazil have traditionally dealt directly with the researcher, bypassing university institutional authorities and structures. This was the procedure followed when graduate-level education was introduced and research capacities were placed in the universities. Funding agencies and other regulatory bodies have changed their approach and now try to take into account university structures, shifting the point of reference to the department. But this policy has caused uneasiness among academics who do research outside the department. "You buy the equipment, and it has to go to the department. But you need it here [at the center] for your research activities and your students—not at the department, where there is no way of knowing who will have access to it" (male, political science, full professor, top tier).

Prestige in the area of basic research has different sources at top-tier and second-tier institutions. In the more competitive departments prestige is based on the quality of the publications of their faculty. Academic power translates into institutional power. "Everyone knows what you are doing and the value of your work. Competition is stiff, but as long as you have the necessary competence, you will succeed" (male, chemistry, full professor, top tier). Academics at these institutions have no problem with evaluation, be it by the university authorities or other agencies. The only cause for concern is the changes in funding patterns for research. In the past, state-owned enterprises were the main source for science and technology investments in Brazil. Work involving these enterprises and academic research teams took the form of large projects, focused mainly on basic knowledge. Privatization disrupted or terminated research programs and the long-established networks for cooperation among universities and state-owned enterprises. Furthermore, funding for collaborative research with the participation of industry—supported by tax incentives for industrial research and development—is outpacing funding for research projects generated solely on the initiative of the academic team (Balbachevsky and Botelho, 1999). These changes have pushed the university research teams to think more strategically about their research agenda. For academics, this imposes the need to learn how to deal with a third party. "In the past, we only had to deal with our peers. Now, we need to cope with the demands from other quarters. We must sell our project and address issues that go beyond the requirements of academic excellence" (male, physics, doctorate, second tier).

In less competitive departments, the bureaucracy is a source of prestige; in some cases, unions have a similar effect. "Here, faculty prestige is enhanced by academic productivity and holding administrative posts" (female, geography, full professor, second tier). Appointment to a committee or to the departmental council, for example, is an important source of power and thus, of prestige. Academics in these departments are uncomfortable with the evaluation process: "They only take numbers into account: how many articles you published, how many books. There is no concern about quality...." (female, geography, full professor, second tier). Academics in such contexts are uneasy about the new criteria used by university authorities in allocating resources. They complain about not receiving what they deserve but do not know how to assert their worthiness.

At third-tier institutions, the role of teacher provides the major source of identity, and authority is derived from the institution's confidence in a faculty member. Each program has a faculty coordinator, and that person reigns supreme: She or he may choose whom to hire, is in charge of evaluations, and communicates directly with the top authorities about the program. It is a vertical structure. For the teacher, popularity among students is also a factor. Here, the measures taken by the Ministry of Education to supervise and evaluate the quality of teaching have had a major impact. Teachers are unsure of how to respond to outside pressures for reform. They see the need for improving the quality of teaching but lack the resources to achieve this improvement. High teaching loads preclude training for a higher degree, and support from the institution is minimal. Moreover, faculty in the third tier are unsure about the function and urgency of the external evaluation they have to undergo. "We don't know what to do. Some colleges that only pretended to make changes get better evaluations than others that have undertaken reform projects. Also, change takes time, but the ministry wants everything done yesterday. This is not possible—not if you want to work with your faculty and improve your colleagues. You cannot upgrade your faculty to the doctoral degree level in just three years or less. It's not so unusual that one school takes the easier path, hiring anyone with a degree, and receives a good evaluation...." (male, sociology, doctorate, third tier, program coordinator).

Conclusion

As stated earlier, the last decade of the 20th century was a time of great changes in Brazilian society. Until the 1990s, Brazil was a closed economy. In the 1990s, the opening up of the economy exposed Brazilian society to an unusually high degree of competition. These changes in turn forced new priorities onto the agenda of the higher education regulatory bodies. Governmental response to this new agenda was expressed in an across-the-board pressure on institutions for accountability and quality control. But the institutional response to this pressure has been very diverse in nature.

The Brazilian higher education system is not only very diversified but highly stratified. The few universities at the top of the system have achieved strong academic standards and have succeeded in developing strong graduate-level programs and in institutionalizing the research enterprise. In part of this system, the academic workplace has undergone little change in the last decade. These universities are strongly insulated from the pressures posed by government and society. External pressures did open some fissures, but this occurred mainly with regard to the external bodies such as the so-called foundations or individual projects. These new developments have not yet achieved full legitimacy within institutions.

The second tier contains the majority of the Brazilian public-sector institutions. However, unlike the best universities, those in this tier failed to develop strong graduate programs and do not have a high proportion of well-qualified academics. Nevertheless, they have been accredited as universities, which means their faculty hold high expectations concerning status and working conditions. Faculty in these institutions have access to full-time contracts, with low teaching loads and good career prospects.

One might expect that institutions in the second tier would be more vulnerable to governmental pressures, but that is not the case. Most share with the best universities the legal status of being public universities. As in public universities in the first tier, a number of restrictions inhibit government action. For this sector, any move that appears to acknowledge differentiation is a threat. Until now, faculty, union, and bureaucratic forces have managed to oppose changes in the status quo. As a result, the academic workplace in this environment has also experienced little change. Even when university authorities are aware of the changing environment, they lack the tools to implement the reforms. Thus, in the view of the university's authorities, strengthening bureaucratic controls was the only way to deal with the new challenges posed by the external environment. As they succeed in imposing these controls, more and more dynamism is lost by the public universities in Brazil, whether in the first or second tier.

At the bottom of the system are a great number of large and small institutions highly responsive to the demands of the job market. For these institutions, the language of the academy is a foreign one, although they are being forced to start learning this language through governmental pressure. The third tier is the most vulnerable to pressures from the government's regulatory agencies. Thus, it is in this tier that the changes have been the most dramatic. One of the more noticeable changes involves an increase in the proportion of full-time contracts and qualified academics among the faculty of institutions in this tier. Nevertheless, the actual impact of such changes at the institutional level is hard to assess. As stated above, all Brazilian higher education institutions have an ingrained pattern of evasive, bureaucratic responses to external pressures. It is too early to tell how much of this change consists of made-up figures, and how much is real.

Regardless of its relative resistance to change, the higher education system has been experiencing growing external pressures in the direction of role diversification and accountability in recent years. Success in reforming the Brazilian higher education system should result in a more inclusive, differentiated, and less hierarchical framework. One could thus conclude that the system's ability thus far to resist the pressures toward the differentiation and marketization of higher education in Brazil has not created a brighter long-term perspective for the Brazilian academic workplace.

NOTES

1. High inflation rates have more than three decades of history in Brazil. Rates above 100 percent per year have been registered since the late 1980s. At the beginning of 1994, when the "Plano Real" stabilization program was launched, Brazil's inflation rate had reached more than 1,000 percent per year. This long period with high inflation rates created a peculiar business culture in Brazil, where profits were to be sought in the overnight financial market rather than in good management. This, in turn, had an important impact on the demands for qualification and training offered by the higher education system.
2. The great majority of such institutions are family-owned enterprises.
3. In Brazil, the prevailing notion is that the primary role of higher education is to provide training and certification for established professions. Credentials are a necessary condition for access to protected job-market niches. This is not an unusual concept of the role of higher education worldwide, but what is unique in the Brazilian case is the broad application of this idea. Here, almost every graduate diploma is associated with a protected labor market niche.
4. Beside the São Paulo state system, there are state-owned institutions in 15 other Brazilian states. All have autonomy from federal supervision, but the São Paulo system also has financial autonomy.
5. Brazilian research funding agencies operate largely through programs supporting research projects instead of through institutional support. The only exceptions are the funds administered by CAPES. But even these are directed toward graduate programs, bypassing all other university levels and the bureaucracy.
6. One of the most prominent of these is the Universidade de Mogi das Cruzes. A large, private, for-profit university situated in a town near the city of São Paulo, it hired a former rector from the University of São Paulo, nominated him as rector and started an impressive reform program. In 2000, the genome research group created at this university was accepted as a member of the State of São Paulo genome network—a group of centers and laboratories supported by the State of São Paulo Science Foundation.
7. This section is partially based on the arguments presented in Balbachevsky, 2000.
8. Brazil was under military rule from 1964 to 1984. Liberalization started in 1974, and in 1984 a democratic government was elected for the first time since 1964.

9. Before the reforms, workers in Brazil had the right to retire after working 35 years, for men, and 30 years, for women. For certain professions, including teachers and academic professors, the time span required was shorter—30 years for men and 25 for women. The reforms linked this requirement with a minimum age of 65 for men and 60 for women.

REFERENCES

Balbachevsky, E. (1999). *A profissão acadêmica no Brasil: as múltiplas facetas de nosso sistema de ensino superior* [Academic profession in Brazil: A portrait of the multiple facets of our higher education system]. Brasilia, Brazil: FUNADESP.

Balbachevsky, E. (2000). From encirclement to globalization: Evolving patterns of higher education in Brazil. In M. S. McMullen, J. E. Mauch, and B. Donnorummo, *The emerging markets and higher education: Development and sustainability* (pp. 149–170). New York: RoutledgeFalmer.

Balbachevsky, E., and A. J. J. Botelho. (1999). *Marcos e desafios da política científica e tecnológica* no Brasil [Milestones and challenges of Brazilian science and technology policies]. In J. Bellavista and V. Renobell (Eds.), *Ciencia, tecnología e innovación en America Latina* (pp. 117–148). Barcelona: Publicaciones universitat de Barcelona.

Castro, C. M. (1985). *O que está acontecendo com a educação no Brasil* [What is happening with Brazilian education]. In E. Bacha and H. S. Klein (Eds.), *A transição incompleta* (pp. 103–162). São Paulo: Brazil: Ed. Paz e Terra.

Coelho, C. E. (1988). *A sinecura acadêmica: A ética universitária em questão* [Academic sinecure: University ethics in question]. São Paulo, Brazil: Editora Vértice, Ed. dos Tribunais.

Durhan, E. R. (1998). *Uma política para o ensino superior brasileiro: diagnóstico e proposta* [Politics for the Brazilian higher education: Diagnosis and proposal]. Sao Paulo, Brazil: Universidade de São Paulo.

Klein, L. (1992). *Política e políticas de ensino superior no Brasil: 1970–1990* [Brazilian higher education politics and policies: 1970–1990]. São Paulo: Universidade de São Paulo.

MEC/INEP (1998). *Sinopse do Ensino Superior* [Brazilian higher education synopsis]. Brasilia: MEC.

Neave, G. (1993). Séparation de Corps: The training of advanced students and the organization of research in France. In B. R. Clark, *The research foundation of graduate education: Germany, Britain, France, United States and Japan*. Berkeley: University of California Press.

Oliveira, J. B. A. (1984). *Ilhas de competência: carreiras científicas no Brasil* [Isles of competence: Scientific carreers in Brazil]. São Paulo, Brazil: Ed. Brasiliense.

Sampaio, H. (2000). *Ensino superior no Brasil: o setor privado* [Higher education in Brazil: The private sector]. São Paulo, Brazil: Hucitec/Fapesp.

Schwartzman, J. (1993). *Universidades federais no Brasil: Uma avaliação de suas trajetórias—décadas de 70 e 80* [Federal universities in Brazil: An assessment of their trajectories, during the 1970s and 1980s]. São Paulo: Universidade de São Paulo.

Schwartzman, S. (1992). Brazil. In B. R. Clark and G. Neave (Eds.), *The encyclopedia of higher education* (pp. 82–92). Oxford: Pergamon Press.

Schwartzman, S., and E. Balbachevsky. (1996). The academic profession in Brazil. In P. G. Altbach (Ed.), *The international academic profession: Portraits of fourteen countries* (pp. 231–279). Princeton, NJ: Carnegie Foundation for the Advancement of Teaching.

Velloso, J. (1987, May). Política educacional e recursos para o ensino: O salário-educação e a universidade federal [Educational policy and finance resources to learning: The "salary-education" program and the federal university]. *Caderno de Pesquisa*, 3–28.

Velloso, J., and L. Velho. (2001). *Mestrandos e doutorando no país: Trajetórias de Formação* [Master's and doctoral students in the country: Training trajectories]. Brasilia: CAPES.

5

THE ACADEMIC PROFESSION IN CHINA

Xiangming Chen

As the world enters the 21st century, the academic workplace everywhere is undergoing changes in its development, operation, and status. This chapter examines the Chinese academic profession at a time of rapid transition from bureaucratic control to market forces, and the country's shift from a command economy to a market economy. Data from in-depth interviews, documents, and case studies will be used to outline the major changes, analyze the evolution and consequences of the process, and point to the challenges for the Chinese academic profession.[1] The focus will be on the socioeconomic context of these changes, how the changes (heavily influenced by Western practice) relate to Chinese cultural beliefs and norms, and the implications for the academic profession. The analysis will examine the impact of the introduction of a "post system" on the existing "title system" in academic appointments—a reflection of the change from a predominantly bureaucratic to a competitive corporate culture in China.

The study concludes that the Chinese academic profession is still caught between the pull of the old bureaucratic controls and the push of market forces. The introduction of market mechanisms has brought about greater academic freedom, attention to quality, and transparency of promotions in the academic profession, but the old forces of central planning and official interference are still at work. The Chinese academic profession faces the challenge of deciding what to keep and what to discard from its traditions, and what to adopt and modify from the world trends of modernity.

Modern Chinese higher education has developed rapidly in the past 100 years, since its establishment in 1898. At present, there are 1,942 state-owned higher learning institutions, among which are 1,071 regular higher learning institutions and 871 adult institutions.

Table 5.1 Faculty and staff at Chinese public higher learning
institutions, by job category, 1999

Category	Number (thousands)	Percentage
Full-time faculty	42.7	39.97
Support staff	131.50	12.35
Administrators	179.6	16.87
Technical staff	144.20	13.54
Research staff	49.50	4.65
Enterprise staff	57.80	5.42
Staff of affiliated institutions	76.7	7.20
Total	1,065	100.00

Note: From Ministry of Education, 2000.

In 1999, the total number of full-time faculty and staff in the public
sector was 1.065 million (see table 5.1 for distribution by job category), and in 2001, student enrollments were about 10 million. Private
institutions, a newly emerging phenomenon, were reportedly over
2,000 in number as of 2001, with a lower level of student enrollments
and a less stable faculty.

QUANTITATIVE ISSUES

During the period from 1949 to the 1990s, Chinese higher learning
institutions basically employed an "all-tenure" system in academic
appointments. Once hired, faculty and staff were provided with an
"iron rice bowl" (i.e., a secure job for life), and all were paid more or
less the same, regardless of their qualifications and contributions. This
rigid system has resulted in a serious problem of overstaffing in almost
all higher learning institutions, except those in the private sector.[2]
In 1998, the ratio of students to faculty and staff combined in the
public sector was 5.1 : 1, and the students to faculty ratio was 11.2 : 1
(Yao and Gong, 1999). As these figures indicate, the overstaffing is
a problem more with administrative and supporting staff than with
faculty.

The overstaffing problem has grown worse with the mergers of
higher learning institutions and the resulting overlapping administrative offices. The purpose of the mergers was to produce economies of
scale, to delegate more administrative power to the local authorities,
and to make institutions more comprehensive. Since 1992, 179 institutions have been merged, and the student enrollments at each

institution increased on average from 2,380 in 1995 to 3,500 in 2000. A greater variety of disciplines and curricula are offered to students at each institution, and more resources are shared among the merged institutions. Nevertheless, the number of faculty and staff are multiplied by the number of the merged institutions, although some posts are eliminated with their corresponding staff and some faculty are fired or reassigned. If a smaller, less-renowned institution merges with a larger, prestigious one, the new entity usually takes the name of the latter as well as its standards for faculty appointments. Consequently, many faculty members from the smaller institutions are placed at a disadvantage because of their level of competence. After Central Technical Arts College was merged with Tsinghua University in 2000, for example, only 6 of the 12 English teachers at the former were retained in the English Department of Tsinghua University. The rest were forced to find work at other institutions or in other professions. For this reason, faculty from smaller institutions are generally not very enthusiastic about mergers, while their students are very happy.

While the overstaffing of administrators is a general problem nationally, almost all institutions suffer from understaffing of competent faculty, due to the recent expansion of higher education. In the past ten years, student enrollments in China have risen to meet the increasing need for higher education (see table 5.2). The number of first-year students at regular higher learning institutions increased 47.3 percent, from 1.084 million in 1998 to 1.597 million in 1999. Enrollments at adult learning institutions increased 15.7 percent, from 1.001 million to 1.158 million, and the number of graduate students increased 17.6 percent. Total student enrollments doubled between 1990 and 1999. The gross enrollment rate of the 18–22 age cohort increased from 3.9 percent in 1992 to 9.1 percent in 1997, to 10 percent in 2000, and is expected to rise to 15 percent in 2005.

Table 5.2 Expansion of higher education enrollments and institutions, 1990s (thousands)

	1990	1995	1998	1999
Total	3,822	5,621	6,430	7,423
Graduate students	93	145	199	234
Undergraduates	3,729	5,476	6,231	7,189
At regular institutions	2,063	2,906	3,409	4,134
At adult education institutions	1,666	2,570	2,822	3,055

Note: From Ministry of Education, 1990, 1995, 1998, 1999.

Given this rapid expansion, most institutions have more students than they can handle, not only in terms of services (accommodation, classrooms, laboratories, library space, and books) but also in terms of the number of faculty. The faculty workload is increasing, as is class size. One survey indicates that each teacher is advising 26 master's students and that classes with 200 to 500 students are not uncommon (Miao, 2001). As most higher learning institutions in China have not yet introduced the system of having teaching assistants to lead discussion sections, the quality of teaching is reportedly deteriorating. The predominant teaching method is still "duck feeding," with the teacher lecturing to a large audience, whose main task is to take notes passively. Personal interaction between teachers and students is minimal, and students often complain about the lack of guidance from teachers.

The problem of understaffing is especially serious at local vocational and adult institutions, which have experienced the most expansion. Having to endure relatively poor living conditions and salaries and lacking opportunities for professional development, many qualified and competent faculty members at these institutions are dissatisfied with their jobs and tend to leave as soon as possible.[3] New graduates from universities also prefer not to work at such institutions, because of their poor conditions.

> *Case 1:* Located in a small city in Province A, College A is an institution under the governance of the province, the city, and the factory that started the college in 1972. In 2001, student enrollments at this college increased by 33 percent, but the number of full-time faculty decreased by 15 percent. Graduates with bachelor's degrees are not qualified to teach, and no master's degree holders would take up posts at the college. Before the expansion, some teachers could moonlight to get extra income, but now the college does not allow them to take on outside work and assigns them heavier teaching loads for less pay than they would earn from moonlighting. Some teachers are not motivated to teach, not only because of the low pay but because promotion depends less on teaching than on research. Lacking enough support from the college to do research, teachers wishing to develop themselves professionally seek further training, after which some choose not to return to the college. The college is then forced to hire engineers and technicians from nearby research centers to teach, thereby raising personnel costs at an already impoverished college.

Quality Issues

As alluded to earlier, the biggest problem facing the academic profession in China is not so much one of quantity as of quality.

Overstaffing does not mean that there are enough qualified faculty at all institutions. On the contrary, many institutions, especially those located in small cities with weak economies, do not have enough qualified or competent faculty. Research universities affiliated with the Ministry of Education or other ministries do not have difficulty attracting well-known, competent people. However, because of the remnants of a rigid personnel system, it is not easy to get rid of incompetent people.

The Teachers' Qualification Law, enacted in 1995 by the State Council, stipulates that the minimum requirement for faculty below the age of 40 is a master's degree. However, the qualification levels of faculty are still rather low, although graduate programs have mushroomed in recent years. By 1999, only 5.4 percent of faculty held a doctoral degree; 24 percent had earned a master's degree (see table 5.3 for distribution by title and by educational background).[4] Even though in recent years more and more Ph.D. holders have returned from foreign countries, the return rate is still rather low (reportedly 2 percent), and the system is still suffering from serious brain drain to the more developed countries.

The data further indicate that the age distribution of faculty is rather problematic. Although the situation has been improving since the early 1980s—immediately after the Cultural Revolution, which deprived a generation of higher learning—the age structure is still not

Table 5.3 Full-time faculty in Chinese public higher learning institutions by title and education, 1999

Education	Professor	Associate professor	Lecturer	Assistant teacher	Teacher	Total
Ph.D	6,331	10,500	5,351	508	446	23,136
M.A. degree	7,158	30,864	44,339	13,964	4,167	100,492
Failed to get graduate degree	1,577	2,631	3,156	962	153	8,479
B.A. degree	10,113	47,390	84,811	61,314	14,066	217,694
Dropped out of M.A. program	107	250	361	113	16	847
Failed to get B.A.	13,187	27,444	10,499	2,970	879	54,979
College diploma or 2 yrs. college	791	6,391	7,114	3,061	988	18,345
Less than 2 yrs. college	95	430	759	304	122	1,710
Total	39,359	125,900	156,390	83,196	20,837	425,682

Note: Ministry of Education, 2000.

well balanced for sustainable development. There are too few competent middle-aged faculty to fill the gap between the young and the old (Jiao, 1998). In 1999, 47 percent (201,669) of all faculty (425,682) were between the ages of 36 and 50; but of these, only 5.57 percent (11,237) were full professors. Of the 39,359 full professors, 23.33 percent are already over the age of 60 (Ministry of Education, 2000).

One of the major problems regarding quality is the rigid personnel system, which has prevented the entry of competent people into the academic profession and the departure of incompetent people. Under the pre-1980s command economy, faculty employment basically followed the top-down assignment system for appointments and the seniority system for promotions. The "iron rice bowl" syndrome was as predominant among university faculty as in any other profession in China. Faculty had little incentive to work hard because of the uniformly low salaries and lifetime job security. Institutions had neither the autonomy nor flexibility to hire appropriate faculty to meet specific needs, causing overstaffing or ill-suited staffing. Since salaries were relatively low, a number of teachers, especially younger and middle-aged ones, chose to leave the academic profession for more lucrative ones.

Another cause for the problems concerning faculty quality is inbreeding. Older faculty and administrators like to employ their own students. Thus many universities, especially the prestigious research universities, employ their own graduates. As a result, the ways of thinking and operating are rather similar among different generations of scholars. The absence of new blood from the outside only adds to the lack of mobility among faculty.

RECENT CHANGES IN ACADEMIC APPOINTMENTS

China is now in transition from a socialist country with a command economy to a socialist country that is integrating elements of a market economy into the system. In order to become competitive by international standards, China is now striving to develop itself into a political and economic power internationally. Thus, education, especially higher education, has been targeted as a major means of improving the qualifications of the citizenry and of creating the new knowledge and skills required to empower the nation. As a result, raising the quality of university faculty has been put at the top of the national agenda, and many reform strategies have been introduced in faculty development.

Reform is inevitable also because of the current national and international trends in the academic workplace. The old employment system has proven ineffective in attracting competent people, motivating faculty to work hard, or allowing the sharing of human resources among different institutions. A more competitive system for excellence, quality control, and mobility of faculty is gradually being implemented in Chinese higher learning institutions.

The Post System

The major reform introduced in the 1990s has had the effect of shifting the priority in academic employment from seniority to quality—that is, from the "title system" to the "post system." Under the title system, promotion is decided according to length of service, while the post system depends on the match between the requirements of a post in terms of clearly defined qualifications and responsibilities, and the actual strengths and abilities of faculty. The post system allows for appointments at certain ranks with corresponding budget allotments, while the former title system is tied to fixed salaries.

In 1998, Peking University and Tsinghua University were singled out by the Ministry of Education as experimental sites to become premier world-class universities in the near future. In 2000, each university was awarded 1.8 billion yuan (U.S.$222 million) annually for three years from the ministry, with 25 percent of the funding designated for faculty development. The two universities have set up nine ranks with three categories (A, B, C), each category containing three ranks. Subsidy for the ranks is predetermined by the university, with the highest rank receiving 50,000 yuan (U.S.$6,200) a year and the lowest rank 3,000 yuan (U.S.$370) a year, creating a 17-fold difference. Above category A, there is another special rank, with a subsidy of 300,000 to 500,000 yuan a year for internationally known scholars (Chen, 2000). Only 70 percent of the subsidy is paid to the faculty member each month; the remaining 30 percent is paid after the faculty review at the end of each year.

This reform has the following main objectives: (1) to break away from the "all-tenure" system with its "iron rice bowl syndrome"; (2) to separate one's title from one's rank so that competent people can be hired for the right posts even if they do not have a certain title; (3) to link one's salary with one's post, thus widening the gap between different salary scales; and (4) to downsize the faculty by reassigning surplus people to other jobs.

Under this system, if they are not competent enough, even full professors may not be selected for the category they desire (usually category A). Consequently, they are not entitled to the corresponding subsidy, over and above their basic title salary, which at about 1,000 yuan (U.S.$120) a month is hardly sufficient to support their daily living expenses. In 2000, 65 full professors (one-ninth of their total number) at Peking University were not approved for category A, whereas 71 young associate professors were approved for category A. At Tsinghua University, only 60 percent of faculty and staff qualified for the post subsidy, with all the rest receiving only the basic title salary (Gong, 2000). One of the purposes for setting up three ranks in each category is to prevent tenured professors from getting too "relaxed" to work hard, especially the younger full professors.

The standards for setting up the ranks relate to the employee's qualifications and competence, as well as to the requirements of the post. The procedure combines both top-down and bottom-up approaches. The university may design its own three- or five-year faculty development plan. Each school and department may also submit its own plan to the university for approval. The personnel department of the university designates a quota for each school and department. The criteria are student enrollments; number of key state laboratories, research centers, and projects; and the importance of the fields of study—both long-standing and newly emerging disciplines. If a school or department has a large student enrollment, or more of the centers and projects mentioned above, it will get a larger quota for posts.

The usual procedure is that each faculty member applies to the academic committee of the school or the department for a certain rank, at the end of each year. The committee then submits its decision to the university committee for approval. Once promotion to the rank is approved, the faculty member signs a one-year university contract that clearly states his or her responsibilities. At the end of the year, the faculty member will write a report about work accomplished during the year, and again apply for a rank for the following year. Those who do not fulfill the requirements of the contract will be dismissed from the post. At Tsinghua University, lecturers and assistant teachers who are not promoted after being considered three and two times, respectively, will not be retained in their academic positions. There is a connection between the types of courses and teachers' ranks. Those who teach "classical courses" are hired in category A, and those who teach "key courses" are paid an additional 5,000 yuan per semester. A special financial award is offered to full professors who teach undergraduate courses. One "teaching incident"—such as absence from class or another infraction—would lead to withholding of the financial award. If a teacher were rated as "unqualified" after three in-class observations carried out by the university's teaching guidance center, he or she would be dismissed from the post. In 2000, one full professor and two associate professors at Tsinghua University were reportedly dismissed, mainly as a result of negative student evaluations because of "poor performance" in teaching (Li, 2001).

Other universities are following the example of Peking University and Tsinghua University in implementing this major reform. In 2000, Fudan University and Shanghai Transportation University, for example, each received a 1.2 billion yuan (U.S.$150 million) joint funding grant from the Ministry of Education and the Shanghai municipal government. One-third of the fund has been used for faculty development, under the post system described above. It is reported that the top professors at Shanghai Transportation University may receive as much as 200,000 yuan (U.S.$2,500) a year (Lei, 1999). At Shandong Agriculture University, the title system has been abolished, and 124 Communist Party officials and administrators officials lost their academic titles in 2000. About 37 percent of full professors and 12 percent of associate professors were not selected for any posts. As a result, the expenses for personnel declined from 94.6 percent of the university's total budget in 1998 to 70 percent in 2000. All savings were used to increase the support for faculty with posts (Yang, 2000).

The new post system signifies a major change in the Chinese academic profession since 1949 and has proven much more effective in motivating faculty to work hard than the previous title system. The rank-related subsidy has become an indispensable part of faculty income, since basic title-based salaries are very low. With the rapid reforms in housing, medical care, and pensions, cost-of-living expenses for professors are much higher than before. Faculty have to find ways to earn extra income in order to maintain a decent standard of living. Before the post system was instituted, many professors were forced to take extra jobs outside the university. Now they are not so eager to engage in moonlighting and are able to devote more time to their own professional development and their commitment to their institution.

The introduction of the post system launched a massive downsizing of faculty and staff in higher learning institutions, as in all government departments and state-owned enterprises in China. Those who are downsized are mostly administrators, Party officials, supporting staff, and a small number of incompetent faculty members. At Xiamen University, for example, the introduction of the post system led to a reduction from 31 to 24 administrative departments and from 54 to 32 academic sections, with a cutback in the total number of staff from 357 to 270 (He, 2000). At Central China University of Science and Technology, 290 staff members were cut from the total of 450 in 2000.

The biggest headache for the university leadership is to find ways to reassign surplus people. According to state policy, faculty and staff from higher learning institutions may not be dismissed into society. They have to be reassigned to other posts within the same institution, especially for the first year. Many methods have been invented to accommodate these people (Chen, 1999). Staff may be moved over to university services or affiliated enterprises—such as libraries, dining facilities, real estate, kindergartens, and primary and middle schools. Early retirement is an option—at age 45 for females and age 50 for males. Where appropriate, a faculty member may go on sick leave while still enjoying all the benefits from the institution. Some individuals obtain further training at a higher learning institution (usually in the case of people below age 30), with a view to finding new jobs later on. Some faculty find jobs outside the university, but keep their institutional affiliation, with benefits, by paying a certain percentage of their income to their institution. Others find new jobs elsewhere and leave the institution. As a last resort, faculty may send their resumes to the institution's personnel exchange center or that of the local community for job placement, while receiving an allowance from the municipal welfare system for a fixed period of time. Most faculty who are removed from their original posts are able to find new jobs, although some express dissatisfaction with the type of work, salary, or workload.

Competing for Talent

Higher learning institutions are concentrating more on attracting competent people, while getting rid of incompetent ones. Using financial subsidies as attractions, the Ministry of Education and other ministries as well as institutions have carried out a number of recruitment campaigns. In the past few years, quite a few government programs have been set up to award competent faculty and to attract outstanding scholars to work in higher education—such as the State Outstanding Young Scientists Foundation, Cross-Century Excellent Talent Training Program, Excellent Young Teachers' Support Fund, Overseas Returned Scholars Research Initiation Fund, Key Research Fund for Outstanding Overseas Returned Scholars, and Talent Training Fund.

Some of the programs include generous financial awards, by Chinese standards, for the winners. The Excellent Young Teachers' Teaching and Research Reward Fund for Higher Learning Institutions, set up in 1999, provides an annual award of 100,000 yuan (U.S.$1,200) each to 100 young faculty under the age of 35 for a five-year period.

The Long River Scholars Award Program, established in 1998 by the Ministry of Education in conjunction with Li Jiacheng, a Hong Kong businessman, has created 365 professorships in 258 universities, with an annual subsidy of 100,000 yuan, plus other benefits provided by the state (Jiang, 2000).

Another strategy to retain competent faculty members is to provide them with the opportunity for advanced study and research. Many new programs have been set up for faculty to get in-service degree and diploma training (e.g., Domestic Visiting Scholars Program, Senior Seminars, and Master's Degree Course Program). At Shanghai 2nd Medical University, for example, the university set up a model program, called 3-T (i.e., top talents, top projects, and top scientific achievement). Faculty members from different departments are grouped together while studying abroad, so as to encourage interdisciplinary learning and to ensure a higher return rate. In 2000, 150 faculty members below 40 years of age were singled out as academic successors of the older generation, each enrolling in a two-year in-service training program. Each year the university has also invited overseas Chinese scholars to teach and do research for six months, providing them with first-class housing, salary, subsidy, laboratory facilities, and assistants.

While prestigious universities may send their faculty out for further study, small local institutions are unlikely to follow this strategy to retain competent faculty. While these institutions also want their faculty to develop professionally, they are incapable of providing the necessary training locally and fear that if they let faculty go elsewhere for further training, they will lose them for good.

A number of preventative measures are used by the ministry and by institutions to ensure that faculty return to their institutions after further training. One method is to ask the person leaving for training to provide a colleague to act as guarantor. For a Ministry of Education scholarship program, the insurance required is 50,000 yuan, about U.S.$6,200. If the faculty member does not return on time, the guarantor would be held responsible for the insurance fee, and the money would not be refunded. Actually, it often occurs that not only does the faculty member not return, but the guarantor also leaves the institution to evade the punishment. Those who go abroad and do not return can usually afford to pay back the insurance money to their guarantors, since scholarships from foreign institutions are usually quite large by Chinese standards.

Currently, some institutions are also using positive strategies to ensure that faculty come back—paying them their regular salaries and granting them promotions while they are away. If they return on time, they receive an additional financial award, in addition to the return of the insurance money. One local college even pays a faculty member's post subsidy when he or she is studying in a doctoral program at a well-known university (Linghu, 2001).

The Contract System

Compared to the public sector, academic employment in the private sector is organized much more simply. A newly emerging phenomenon, since the early 1990s, the private sector has mushroomed to about 2,000 institutions that mostly provide education in a few practical fields of study. The Ministry of Education has specific requirements and procedures for approval of the establishment of a private higher learning institution, but no hard and fast rules about the number, title, or rank distribution, or salary of the faculty. Private institutions are financed mainly by nongovernmental sources, and faculty salaries are much higher than those in the public sector. In addition to providing an alternative channel to meet the increasing demand for higher education, the private sector also offers a different model for faculty appointments.

With few exceptions, private institutions operate on the contract system for faculty appointments. Teachers are usually hired on one-year contracts or by the term. At some institutions, full-time teachers are hired on monthly contracts, and part-time teachers are paid by the hour. At others, all teachers are paid by the hour. Private institutions usually do not provide benefits for their faculty.

In general, private institutions have only a very small proportion of full-time faculty. Full-time faculty are mainly professors retired or transferred from public universities, or new graduates. They are extremely busy with multiple duties as teachers, secretaries, internship tutors, guidance counselors, class managers, accountants, and so on. Part-time teachers are responsible solely for teaching. The teacher/student ratio is generally much higher than in the public sector. At Beijing Haidan University, for example, the ratio is 1 : 54.[5]

Except for a few institutions accredited by the ministry, like Beijing Haidan University, most private institutions do not provide their faculty with the opportunity for promotion—in any case the number of full-time faculty is very small. The few institutions that are entitled to offer promotions must apply to their district personnel bureau. There is no quota for promotions in each institution, and the criteria are academic qualifications and performance.

Faculty appointments, payment levels, and promotions in private institutions are determined by graduation rates and by student examination passing rates in the annual national self-exams and institutional exams. Out of all the 112 private institutions in Beijing, only 24 institutions are qualified to refer students for the national self-administered exams.[6] Of the over 2,000 private institutions nationwide, only about 20 are entitled to award diplomas. No private institutions are as yet qualified to offer academic degrees.

THE CHALLENGES

Reform strategies in the academic profession over the past few years have definitely improved the efficiency and effectiveness of both teaching and research. The criteria and procedures for faculty appointments and promotions have become fairer and more transparent than before. With the reforms, while some old problems are being solved or dealt with, new challenges are being revealed in the process. In order to keep the reforms going to their maximum effect, we need to face these challenges with a reflective mind and think through the cultural, political, social, economic, and academic implications for Chinese higher education in particular and China's socioeconomic development in general.

Tensions among Faculty and Institutions

The post system in academic appointments has caused all kinds of tensions among different parties, mainly due to the "iron rice bowl" mentality of the last 50 years. The current reform is in essence a redistribution of interests, resources, and power, as well as a reorganizing of relationships and structures. Those who have lost their advantages will naturally feel deprived by the reforms and harbor a sense of unfairness, resentment, and uncertainty.

First, the reform has created tensions among faculty members, especially between the more competent and the less competent. In the past everybody received the same low payment, whereas now the differences might be as much as 17-fold. Faculty who are assigned to higher ranks with higher salaries may feel under pressure and obliged to work harder, while those who receive lower subsidies may not be prepared psychologically to cooperate with colleagues of higher rank.

A certain amount of tension also exists between younger and senior faculty. Promotion has become more difficult for senior faculty. In the past, they needed only to wait for the years to pass, but now they are required to work hard to gain the necessary credentials, as well as to wait for vacancies. For faculty under the age of 40, their chances are much better than before, since they can get promoted more quickly by "jumping the queue" or by means of a special permit from above for their outstanding achievements. In some cases, a corrupt administrator might use the option of accelerated promotion to advance his or her favorite candidate, who may not be competent for the post, causing more conflict with the older faculty. Although the problem is not widespread, it could do damage to efforts to control faculty quality. The old seniority system could at least ensure that the promoted teachers had the right number of years of experience, and in many ways competence does relate to experience

New tensions have also arisen among different institutions. In general, the prestigious research universities, such as Peking University and Tsinghua University, have the competitive advantage in the national contest for academic talent. Schemes like the World First-Class University Project and Establishing 100 First-Class Universities by 2000, sponsored by the Ministry of Education, add to the existing status, influence, and attraction of these top institutions. Competent people at small local institutions tend to flow to the prestigious institutions, further widening the gap between the haves and the have-nots. In the past, the national competition for the best brains took place mainly between the education sector, government agencies, and other professions—like business. Now, with the intervention of government policies, among other factors, the competition is also fierce among higher learning institutions. In its current drive to evolve from a technological university into a comprehensive university, Tsinghua University, for instance, has succeeded in attracting many outstanding scholars in the humanities and social sciences. It is reported that scholars ranked among the top ten nationally in their fields can get hired right away at Tsinghua and be provided with good living and working conditions (Chen, 1999; Zeng, 1999).

A question then arises concerning the cultural significance for the Chinese academic profession of all these tensions. Given the fact that the reforms were drawn mainly from Western countries such as the United States, a more basic question would be: "Are the reforms compatible with Chinese cultural norms?" Traditionally, Chinese culture advocates peace, harmony, and contentedness; the current reforms certainly run counter to these values. It is clear that the old system did not work very well in terms of promoting efficiency and social justice, but the post system has also stirred up a lot of tension among different parties. More effective ways need to be found to reform the academic profession in terms of substance rather than just in terms of form—to transform competition into healthy cooperation.

Strategies to reduce tensions have already been initiated at some institutions. Peking University and Tsinghua University have agreed to coordinate their financial incentives by using a similar ranking scheme. Within the same universities, some strategies have also been instituted to improve faculty morale and finances. Some departments at Peking University, for example, have used their own money to compensate low-ranking faculty and staff for losses in income and esteem.

The lingering importance of seniority in the current post system is another conscious strategy to mitigate the tensions. Although the ranks are ostensibly open to those faculty most suited to the posts regardless of title, in actual practice there is still a very strong residue of seniority ranking. Among the 100 A1 rank holders at Peking University, for example, only 84 are actually working in the university. The rest are either retired, on leave, or not doing any teaching or research for health reasons or because of other social commitments. The university's pedagogy research team conducted a small investigation in 2000 and found that in the upper ranks fewer faculty are teaching in the classroom (Li, 2000). With regard to the ranks, faculty are given hints that category A is reserved for full professors, B for associate professors, and C for more junior staff. The set quotas for ranks might be viewed as an effort to mediate potential conflicts between the title system and the post system. In a society that respects age, with its accumulated wisdom and experience, seniority is used as a tool for mitigating tensions in any sphere of life, including academic promotions. In any case, people need time to mellow and enough opportunity to prove their integrity and capacity.

In spite of the tensions, faculty members have already enjoyed some advantages under the new post system. For one thing, the salaries of almost all faculty have increased dramatically, even though the differences among faculty have also grown. As a result, the living standards of faculty have improved a great deal, as is the case in the general population. Almost all faculty members now own their own housing, some (about 5 percent at Peking University) have been able to buy a car, all families possess household appliances such as a television and refrigerator, and some families have a computer and an air conditioner.

Although the newly implemented annual review seems a rather grueling and time-consuming task, some faculty members do report benefits from this exercise. They report that having to write down what they have done in the past year and what they plan to do in the coming year helps them think in a more systematic way about their own professional development. The oral reports are also seen as useful exercises that keep faculty informed about what their colleagues are doing and what lessons they can draw from the experiences of others.

The fight over a limited number of posts may appear cruel and painful to the participants, but some do say that the exercise forces them to think about ways of cooperating with their colleagues. Competition can motivate faculty to work harder and to work together. In a larger sense, it could encourage all faculty and staff to make better contributions collectively, which would benefit not only the institution as a social organization but also each individual faculty member.

Centralization

Although Chinese higher education comprises institutions of various types that are located in regions with different socioeconomic backgrounds and cultural customs, the current reforms in the academic workplace (as well as in other sectors) have more or less followed a similar pattern. All institutions, regardless of type or mission, are bound to implement similar policies, designed by the Ministry of Education and used by the prestigious research universities. This top-down approach with uniform application basically does not take local and institutional contexts into consideration. When small, local teaching-oriented institutions take up a model similar to that used by Peking University and Tsinghua University, the expectations of their faculty for financial gain are raised. However, as these institutions receive very little financial support from the central government, they find it very hard to satisfy their faculty's expectations. They will have to turn to the local government, private entrepreneurs, and institutional enterprises—channels that lack sufficient financial resources.

In setting up the prerequisites for academic appointments, most institutions have adopted the practices of the famous research universities, which are based on faculty academic qualifications and research publications. These criteria are obviously inappropriate for fields, such as Chinese medicine, that require training through informal apprenticeships and accumulated practical experience. Some professional institutions such as Zhejiang College of Chinese Medicine, do take into consideration the work experiences, ideas, and clinical skills of some senior faculty, although this approach is rare (Yuan et al., 1998).

Under the uniform guidelines, all institutions are to review their faculty once a year, and the results of the evaluations are used to determine hiring for the next year. Some faculties complain that overly frequent evaluations are inappropriate for certain fields as well as basic research projects that need a longer timeline than one year. If the one-year term is set uniformly for all fields, some scholars may have to present their research results prematurely. It has been suggested that terms (ranging from one to three years) should be set for evaluating posts, depending on field of study, so that faculty work can be appropriately assessed and the quality of their work guaranteed (Wang, 2000).

The implementation of a uniform model in academic appointments reflects China's legacy of central planning, government control, and lack of autonomy and creativity on the part of institutions in a command economy. Very few initiatives are made from the bottom up, and no serious investigation has been conducted before the central government issues a national policy decision.

One of the manifestations of overcentralization is the fact that most institutions lack the right to promote faculty. Out of the 1,942 public institutions in China, only 125 institutions are authorized to make promotions to full professor; 120 are authorized to promote to the level of associate professor. The rest must apply to the local educational authorities for faculty promotions, and their applications need to be approved by a committee composed of faculty from other institutions. This requirement makes the institutions feel powerless to set their own promotion standards according to their specific contexts. Moreover, faculty from other universities may not be familiar with the context of the particular institution. Since local education authorities usually set very stringent and uniform standards based on publications, competent teachers often have to wait many years before being promoted. They may lose confidence after a few attempts and eventually leave their institutions for other professions or institutions (see case 2, below). In recent years, there has been a call for the establishment of professional organizations for faculty evaluations and promotions, so that faculty can be evaluated by experts in their fields rather than by nonexperts from outside and officials from above.

Case 2: College B is an example of an institution that lacks the right to make faculty promotions. To make promotions to the level of associate professor the college must apply to the provincial education commission, which adheres to a very strict quota. Many teachers remain lecturers after a dozen years—with the result that some of them choose to leave the college. The college has five departments, and one year it lost three department heads and vice heads. Every semester, a few key teachers leave, having given up hope for promotions. Hearing these sad stories convinces new faculty to move on before it is too late. (Li, 2001)

Another example of overcentralization is the system in China of appointing doctoral student advisers, perhaps a unique practice in the world of higher education. In China, doctoral programs have to be approved by the Ministry of Education. Advisers are appointed by the universities; and the title of adviser is higher than that of full professor. The system was set up in the late 1980s, when very few full professors had a Ph.D. degree. In order to control the quality of doctoral programs, the ministry approved just a few doctoral programs nationwide and appointed a small number of full professors as advisers (setting a quota for the alotted number). A problem with this system is that many competent full and associate professors cannot formally take on advisees, even though they may actually be doing the work for the few, mostly elderly, official advisers. Conversely, in some cases doctoral advisers who do not draw enough student applicants may lower their academic standards, or even compete with other advisers, to attract enough students to sustain their own position and reputation. Nowadays, with more faculty acquiring doctoral degrees, there are increasing calls for a freer system, one that gives students more leeway in choosing their own advisers.

Another aspect of the top-down approach is that students' perspectives are hardly considered at all in faculty promotions. Although some classes are evaluated, the results are not effectively used for improving faculty teaching or in academic appointments. "Everything operates inside a black box," was the comment by one student in a public campus on-line communication at Peking University, "No student voices are heard." Many teachers who are highly regarded by their students have not been promoted, because they do not have enough publications or do not know how to "market themselves." Some students have begun to ask questions: Are department heads and other teachers really the most qualified to evaluate teaching performance? Have they sat in our classes? Are they aware of the opinions of students? Have they carefully read our teachers' published works? We students are best qualified to evaluate our teachers, so why are we excluded from the whole process?

A related issue is the interference of officialdom into academia. One manifestation of this is the attempt by administrators at various levels to obtain professorships or research fellowships, even if they do no teaching or research. This practice is seen as a way of having "academic experts" run the university with a minimum of additional resource allocation and disruption. Unfortunately, to add the prestige of an academic title to their administrative power and privileges, some administrators have tried to produce academic papers, in haste and often of poor quality. In the meantime, many academics are taking on administrative roles while keeping their academic titles and posts. Although these faculty members may no longer be engaged in serious scholarly research or teaching, their duties as doctoral student advisers or heads of administrative offices are often neglected due to their many other commitments. These faculty are also depriving junior administrators of opportunities for promotion, with negative effect on the latter's career prospects.

In a number of ways institutional policies have supported the trend of faculty becoming administrators. At Peking University, for example, when new housing becomes available for distribution among faculty and staff, one's title or position may be an advantage. The title of head of an administrative department is considered equivalent to that of full professor, and that of vice head is equal to that of associate professor. Being the head of an administrative department also brings benefits in terms of greater subsidy, power, and personal connections. Therefore, talented young faculty members who are offered administrative positions find it very difficult to resist the temptation. As a result, their talents for research and teaching are lost to the institution, in a sense.

The interference of officialdom into academia also takes the form of nepotism in academic appointments. Administrators may use their power to put their favorite candidates into certain important posts. Thus, competitive selection may be used only for ordinary faculty and not for the appointment of high-ranking leaders in the educational administration (see case 3).

Case 3: After several small schools were merged to form University C in early 2001, the new president and Party secretary declared that every administrative post at the department level would be awarded on the basis of competition, with two exceptions: the head of the Party's Organization Department and director of the General Office of the University. These two posts would be awarded to persons selected by the former Party secretary of the university, who had been transferred to another college. The fact that the two posts were to be given to two favorites of the Party secretary provoked so much talk that the Party committee of the university was forced to open up the competition for these two posts to anybody who wished to apply.

At the same university, Ms. Chen, who has a doctorate in physics, decided to compete for an administrative post in charge of teaching. She had been in charge of teaching at a college for three years and had been regarded as very competent. As part of the process, she was now required to give a public speech outlining what she hoped to accomplish. Having never given such a speech before, she felt very nervous. Her husband—who is also a university professor—expressed his opposition to the new process. He explained, "in principle, I don't agree with the procedures. If competition is to be introduced at the administrative level at all, it should be practiced from top to bottom. The posts held by our premier and the officials in the Ministry of Education should also be awarded on the basis of competition among scholars in education. What's the point of introducing a competition for such minor posts?"

The lesson to be drawn from this case is that piecemeal reforms like the post system that have been introduced in the Chinese academic workplace will not be fully effective without substantial changes in the larger social and political context. Any reform is a systematic and comprehensive process that requires the cooperation of the different parts of the whole system. China is still an officialdom-centered society. Therefore, reform should be carried out not only at the bottom but also at the top, and reforms should not only be initiated from the top but also from the bottom.

Quality Standards

With the new appointments system, the concern for faculty quality has moved to the forefront of institutional management and evaluation. What defines people of "good quality"? What are the criteria and standards for "good quality"? And how is it to be evaluated? Questions of this kind are being asked in institutions and government offices nationwide.

The current practice in China, in keeping with international trends, is to focus evaluation more on research than on teaching. Teaching is hard to evaluate and measure, and the usual standards used by institutions consist of such vague statements as "effective teaching and sufficient teaching load" (Xu, 2000). Since the motto of "publish or perish" also applies in China, faculty themselves pay more attention to research than to teaching. The Ministry of Education has initiated some teaching awards with financial compensation, but their scope is too small to impact the majority of teachers.

In many institutions, faculty quality is evaluated and ranked mainly according to uniform quantitative standards, without differentiation among different disciplines in the same institution (Cui, 1996). To assess research, a number of international citation indexes are checked (e.g., the Science Citation Index, the Social Science Citation Index, and the Arts and Humanities Citation Index). Other considerations in promotion decisions include publication in certain key national journals, attendance at international conferences, as well as awards, research projects, and other criteria.

Assessing teaching involves quantifying teaching load, graduate student advisees, administrative service, and awards. At Beijing Normal University, the best teacher training institution in China, the performance of all teachers is rated on a quantitative scale, and financial awards are given out to faculty for each publication. For example, 5,000 yuan is awarded for an article in a nationally known interdisciplinary journal, 2,500 yuan for a publication in a national first-class journal, and 1,000 yuan for a publication in a second-class journal.

This practice has caused a lot of resentment among faculty members who believe that the academic profession is a combination of art and science and is not easily quantified. In addition, different fields have their own characteristics with regard to productivity and cannot be measured by the same indicators. People in fields like math and computer science may show their talents at an early age, while those in the humanities need a protracted period and years of experience to make true breakthroughs. A well-known mathematician, Yang Le, expressed his concern in the *Beijing Morning Newspaper:* "Overemphasizing SCI [the Science Citation Index] would distort evaluation. An active physicist may publish eight papers a year, and a paper four to five pages in length may have five or six authors, with 30 to 40 citations. However, a mathematician would be doing quite well to produce two papers a year, about a dozen pages long, citing no more than 10 papers many of which may have been written decades earlier. ... Old citations are not counted in China's current evaluation scheme, only those cited in the previous two years" (Zhou and Zhang, 2001). Scholars such as Yang Le point out that making great breakthroughs in basic science research often requires working in obscurity for 10 to 20 years. If the university establishment does not understand this and looks only at what is produced in the short term, the consequences would be very harmful for the development of these fields.

Because the standards for academic evaluation are so explicitly quantitative, faculty members are required to put out visible products very quickly—on an annual basis. As a result, people are forced to learn how to package and sell themselves, especially junior faculty members who are still at the bottom of the academic ladder. They tend to appear restless, aggressive, and boastful—qualities that conflict with the cultural norm of the "good scholar" who, in the traditional Chinese view, should be calm, gentle, modest, and wise in demeanor. However, the current system does not lend itself to such kinds of scholars. Interestingly, some scholars who have visited places like the United States tell others to become more boastful and aggressive in promoting themselves: "In the United States, you say you know 100 percent even if you only know 60 percent; while in China we say we know 60 percent even if we know 100 percent. The Chinese way is not working any more in international competition, and we need to learn how to sell ourselves." Given the pressures of international competition and the domestic drive toward modernization, as defined by the West, China is emphasizing worldwide standards rather than its own ancient cultural norms. In actual practice, academics who know how to sell themselves often get ahead much faster than those who are truly talented but remain quiet and modest. In a sense, the definition and standards of a good scholar, or even of a good or a capable person, have changed so much that many senior scholars find it hard to know what to think and how to act.

Since evaluation is based mainly on research, the quality of teaching is deteriorating, especially after the expansion in enrollments. The Ministry of Education offers a few programs to help some key teachers in their professional development, but the majority of teachers never get a chance to improve their teaching skills. In general, institutions do not provide adequate training in teaching for their faculty. Worse still, many teachers do not feel the need to improve their teaching ability, since teaching is not a priority for promotions.

With regard to professional development, many faculty members complain that they do not get enough institutional support. In a Ministry of Education survey of 600 prominent scholars in Shandong Province, 34 percent of the respondents report that they lack sufficient training to qualify for a new post at the university, 48 percent state that their institutions do not pay enough attention to faculty development, and 47 percent claim that their institutions do not provide faculty with enough training in new knowledge or information technology (Zhang and Li, 2000). If even these prominent scholars are of this opinion, one can only imagine what less successful and less-well-situated faculty would think about this issue.

The contract system introduced in the private sector has also had a negative effect on faculty development. The loose ties binding faculty to their institutions, based on salary alone, have brought about changes in faculty members' sense of affiliation, loyalty, and responsibility. As the hiring and firing process has undergone changes, faculty mobility is now a common phenomenon and sometimes causes instability in institutions. Young teachers tend to take up advanced study while teaching at a private institution, with the intention of leaving once they have their diploma in hand. If offered a more lucrative or promising job elsewhere, they will immediately leave their current institution (Linghu, 2000). If the private institution is not rich enough, it will have difficulty attracting competent faculty for such "hot" fields as computer science and foreign languages. Teachers in these high-demand fields usually expect higher salaries than in fields such as mathematics, philosophy, Chinese, and history.

The contract system in the private sector and the post system in the public sector are driven by the power of money. It is true that money can motivate people to work hard, but too much emphasis on monetary reward rather than on a sense of achievement and satisfaction has proven problematic, especially in the academic profession. First, institutional management feels obliged to give financial payment for anything that faculty do. The inner motivation of faculty is by comparison somewhat neglected. In fact, compared with other professionals, university professors as a group tend to care not so much for financial return as for spiritual satisfaction, so long as they can put food on their table and clothes on their back. They take up this profession mostly because they prefer it, even though others may bring them more material wealth. They choose to work within the university because the institution has a special culture, which is important for their personal fulfillment. If managed well, the academic community can create a sense of belonging and a dedication to excellence, cooperation, and achievement—besides feelings of pride, power, and personal success. The single-minded focus on money may prove a short-lived and shallow motivation for these high-minded people. It is clear that the university is not a factory, an enterprise, or an industry, although it has a service function. Faculty have their own life, and this life needs to grow in a healthy, safe, and stimulating environment. The university has the responsibility to foster this development in order to fulfill its mission of knowledge creation, human resources production, and providing moral leadership to society.

This principle applies to the Chinese context as well. Although it is reasonable for Chinese faculty to demand greater financial compensation for their work, given that they have been mistreated financially in the past 50 years, too sudden a shifting of gears may stifle the intrinsic motivation and spiritual inspiration of Chinese scholars.

Traditionally, Chinese scholarship has been linked to the fate of the nation (literally, the "state family") and the masses (the "big family"). Chinese scholars by nature aspire to work for the good of the community and the world beyond themselves and their families. The individualistic ideology that is manifested in monetary terms in the current reforms is in principle at odds with Chinese cultural beliefs and needs to be modified to suit the Chinese historical, social, and cultural context.

CONCLUSION

The current reforms in academic appointments have many complex implications for the academic workplace in China. The implications are multidimensional, involving cultural traditions, social customs, the political and economic system, and professional development. Many of these issues are being addressed by scholars, institutions, and government agencies, while others seem to be impossible to deal with for the moment.

What is clear is that the Chinese academic profession is at a crossroads, torn between old and new forces—the old traditional bureaucratic control and the new corporate culture.

With the transition from a command economy to a socialist market economy, the Chinese academic profession is undergoing a process of change from fully centralized control to gradual decentralization. Today, Chinese higher learning institutions have been granted greater autonomy, and academics are enjoying more academic freedom than ever before in the period since 1949. The introduction of the title system has put the qualifications and competence of the faculty before seniority in determining academic promotions. With this new system comes a dramatically differentiated salary scale that has put to an end the old lifetime job security, with a uniformly low salary for all. As a result, a corporate culture emphasizing performance, management, efficiency, and accountability has become predominant on university campuses. All these changes reflect a major shift in Chinese academic life from tradition to modernity. The Chinese academic profession is now facing decisions on what to keep and what to discard from its traditions, and what to adopt and what to modify from the worldwide trend of modernity. Chinese academia needs to redefine its missions and roles in an era of conflicting values, interests, and needs.

In spite of the changes, however, the old forces of central planning and official interference are still at work. On the whole, Chinese academic life is burdened with excessive controls from the central government, with its dominant ideology and agenda. There is still not enough autonomy for institutions in faculty hiring and firing. Small, local, and vocational institutions are still not entitled to set up their own employment policies, hiring standards, and salary scales. It has been suggested that the Ministry of Education should introduce policies to allow these institutions to develop according to their own conditions and possibilities. While government surveillance and supervision are necessary, more trust could be given to the "invisible hand" of the market, which works mainly through competition based on quality and institutional reputation.

With regard to hiring, many scholars interviewed for this study expressed the wish that universities abolish the quota system and let schools and departments determine their own faculty structure. If a department decides to undertake pioneering research, it can enroll more high-quality researchers even if it does not have many students enrolled. The government and universities should take measures to protect certain less popular fields—such as, philosophy, history, and the minor languages—by offering high salaries and good working conditions to faculty; while other fields—especially high-demand ones like computer science, law, economics, English language, medicine, and the biological sciences—can be left more open to market competition for faculty.

Faculty selection and promotions can also be assisted by responding to students, the "consumers" of the university marketplace. In the past, the university curriculum was quite rigid and the majority of the courses were compulsory regardless of the quality of the courses or the interests and needs of students. With the introduction of a policy allowing students to select courses under the guidance of faculty, students have become the major arbiters of faculty quality, at least in the area of teaching. If a wider selection of courses were offered to students, who could choose what they liked and needed, those faculty able to meet students' needs would be considered "good teachers."

A more diversified range of criteria and standards for faculty appointments and promotions in different kinds of institutions and disciplines is also expected to evolve. The current more or less uniform pattern reflects the long-term legacy of the socialist system of the past 50 years. The Chinese academic establishment urgently needs to develop a more flexible and adaptive academic evaluation scheme for different institutions and disciplines. At a time when China is striving to achieve a differentiated system of higher education with a variety of types and levels of institutions, taking into account regional and cultural differences, the evaluation system for faculty should also change accordingly. Only then can the academic profession in China, which is now to some extent limiting and suffocating faculty, become more fulfilling and enriching.

Although the current post system has had many positive results, it still needs improving. With the continued influence of seniority and of officialdom, what currently exists resembles a combination of the post system and the title system. Posts should be open to all qualified people, not just to individuals within the same department or the same institution. If the posts are publicized throughout the country, or even internationally, they will attract the most qualified applicants. The process of appointments and promotions should involve more democratic opinion sharing and debate and the transparency and fairness in decision making should be increased.

All in all, many reform strategies have been introduced in the academic profession at Chinese higher learning institutions in recent years, with positive impact nationwide. However, for many historical, cultural, and economic reasons, problems and obstacles remain for all involved. This chapter has examined some of the changes, their positive impact on the Chinese academic profession, and future problems and challenges. If China remains stable socially and economically in the coming years, reforms in the academic workplace, as in other areas of the country, will progress steadily with incremental results, which will gradually lead to the healthy development of the country and its people.

Notes

1. This chapter is based on a study conducted by the Chinese research team headed by the author, with team members Cai Leiluo, Linhu Yanping, and Li Yadong, from September 2000 to June 2001. Data were collected through literature reviews, document analysis, observations, and interviews with 14 people in 8 institutions—including policymakers and implementers as well as faculty members. Gender distribution, age balance, and title structure of the sample were taken into consideration.
2. From this point on, all data concern the public sector of Chinese higher education, unless otherwise specified.
3. In this article, the word "qualification" refers to the educational background; the word "competency" refers to the actual ability to handle one's job.
4. The degree system for graduate study in China was abolished from 1966 to 1976, during the Cultural Revolution, and did not return to normal until the 1980s, and even then on a very small scale. This historical factor has contributed to the fact that many current faculty members over the age of 50 do not have a master's or doctoral degree.

5. The institution's full name is Beijing Haidan On-Foot University. "On-foot" refers to the fact that the institution does not provide housing for its students.

6. These exams are held annually by a special government agency. Students may study on their own or attend coaching classes and then apply to take the exams. Once they pass the required number of exams—a process that usually takes seven to eight years—they will be awarded a diploma or a degree, depending on the field.

REFERENCES

Cai, L. L. (2001). Appointments and promotions of Chinese university faculty. Paper written for this project, Peking University.

Cai, L. L. et al. (2000). Literature review on Chinese university faculty appointments and promotions. paper written for this project, Peking University.

Chen, W. B. (1999). Taking opportunities and facing challenges: To create a new environment for faculty development. *Chinese Higher Education, 22.*

Chen, W. L. (1999). Aiming at a new breakthrough in institutional management reform. *Chinese Higher Education, 3.*

Chen, W. S. (2000). Implementing a post-based hiring and subsidy system in order to deepen reforms of the personnel system in the university. *Chinese Higher Education, 12.*

Cui, J. (1996). Relations between quantity and quality in faculty hiring at higher learning institutions. *Teacher Training Research, 3.*

Gong, Y. S. et al. (2000). Farewell to the old personnel management system. *Chinese Higher Education, 2.*

He, J. K. (2000). Grasping three key links and pushing personnel system reform. *Chinese Higher Education, 17.*

Jiang, Y. J. (2000). Implementing the "Long River Scholar Award Program" to promote personnel management reform in higher learning institutions. *Helongjiang Higher Education Research, 2.*

Jiao, R. (1998). Making use of opportunities to deepen the personnel system reform in higher learning institutions. *Chinese Higher Education, 13–14.*

Lei, M. (1999). Uncertainty and hopes in higher learning institutional reform. *Solidarity Newspaper, 12.*

Li, S. Z. (2000, March 16). Report on an investigation into undergraduate teaching at Peking University. Report presented at a meeting on teaching held at Peking University.

Li, Y. D. (2001). The current situation and future development of Chinese university faculty appointments and promotions. Paper written for this project, Peking University.

Linghu, Y. P. (2000, November 24). Report on an interview with teacher A from Zhongxin Enterprise Management College, Peking University.

———. (2001). The present and future of the Chinese faculty promotion system. paper written for this project, Peking University.

Miao, S. L. (2001, Mach 29). Oral presentation at the doctoral interview at PKU.

Ministry of Education. (1990). *Chinese education statistics yearbook.* Beijing: Ministry of Education.

———. (1995). *Chinese education statistics yearbook.* Beijing: Ministry of Education.

———. (1998). *Chinese education statistics yearbook.* Beijing: Ministry of Education.

———. (1999). *Chinese education development statistics outline.* Beijing: Ministry of Education.

———. (2000). *Statistics of MOE affiliated higher learning institutions.* Beijing: Ministry of Education.

Wang, L. (2000). A brief analysis of the implementation of the post-based hiring system and the protection of the rights of university faculty. *Teacher Training Research,* 1.

Xu, J. F. (2000). An analysis the effect of teaching in senior faculty promotions at higher learning institutions. *Teacher Training Research,* 2.

Yang, H. (2000). Rational post-setting and strengthening scientific faculty assessment. *Chinese Higher Education,* 11.

Yao, F. M., and Gong, Y. S. (1999). Confronting the difficulties and pushing forward institutional management reform. *Chinese Higher Education,* 13–14.

Yuan, Q. et al. (1998). Improving vocational education reform and strengthening faculty development. *Teacher Training Research,* 4.

Zeng, S. Y. (1999). Fifty years of faculty development in China's new higher learning institutions and future prospects. *State Senior Educational Administration Review,* 5.

Zhang, F. L. (1999). The inevitability of hiring high-quality faculty at high salaries. *Chinese Higher Education,* 10.

Zhang, L. X., and Li, G. B. (2000). A study on the environment for professional development at higher learning institutions. *Shandong Medical University Review,* 1.

Zhou, R. P., and Zhang, G. L. (2001, March 20). Chinese scientific research will be at the forefront internationally in just over 10 years: An interview with Yang Le. *Beijing Morning Newspaper.*

6

THE ACADEMIC PROFESSION IN
MALAYSIA AND SINGAPORE: BETWEEN
BUREAUCRATIC AND CORPORATE
CULTURES

Molly N. N. Lee

The academic profession is central to higher education because a successful academic institution depends on a well-qualified, dedicated, and adequately remunerated professoriate. Academics in every society play multiple roles: They teach, carry out research, and provide services to their universities and communities. They are considered "experts" in their fields of specialization and are often called upon to provide expertise to government and industry. In some countries, as public intellectuals and social critics, they are expected to contribute toward the betterment of society. The expectations of the professoriate vary, along with their working conditions. The nature of academic work and the condition of the workplace may differ across countries and over time. The purpose of this chapter is to compare the academic profession in Malaysia and Singapore with respect to recruitment, appointments, and promotions, and to examine changes in the culture of the academic workplace.

As in many other countries, the academic profession in Malaysia and Singapore is well respected and used to attract many of the best brains and most highly qualified people. However, for various reasons, the profession's drawing power has declined. First and foremost, the rapid expansion of higher education in times of economic boom, as witnessed in both countries, means that universities and other institutions of higher learning have to compete with other sectors for high-caliber personnel. The fact that academics are paid as civil servants imposes certain restrictions on their salary scales as compared to those in the private commercial and industrial sectors. The civil service is not at all comparable to the private sector in Malaysia, although it fares better comparatively in Singapore. Many universities have grown large and complex, thus requiring a corps of administrative and technical support staff to ensure the smooth running of these institutions. In line with this development, the bureaucrats have taken over many of the decision-making and policymaking processes from academics, thus diminishing the latter's traditional autonomy and influence on the governance of their own institutions. The state retains its strong control over the universities and continues to demand greater accountability and productivity from academics. The move toward the privatization and corporatization of higher education has introduced a corporate managerial culture into universities—subjecting academics to performance-based evaluations, merit-pay schemes, as well as competition for resources and research funds. This chapter examines the changes that have occurred in the working conditions of the academic profession in Malaysia and Singapore.

CONTEXTUAL BACKGROUND

Singapore, which was part of Malaysia when the latter was formed in 1963, became an independent city-state in 1965. Both Singapore and Malaysia are among the newly industrialized economies in the Southeast Asia region with a per capita gross national income in 1999 of U.S.$24,620 and U.S.$3,390, respectively (World Bank, 2001). When compared to the other countries featured in this book, each of these two countries has a relatively small population—just under three million in Singapore and around 22 million in Malaysia. Thus the academic systems in both these countries are comparatively small—about 20,000 in Malaysia and 7,500 in Singapore.

The higher education systems in Singapore and Malaysia exhibit a number of similar characteristics along with some distinctive differences. Both countries started with two-tier, state-managed and -financed higher education systems. The first tier consists of the universities, whose primary function is to meet the country's high-level labor force needs; the second tier, of polytechnics and training institutes, whose function is to provide midlevel workers with technical skills. In Singapore, higher education is dominated by the public sector, with the private sector playing a peripheral role. But in Malaysia, since the 1990s, the private sector has rapidly become a key player in higher education. Another distinctive difference between the two higher education systems is the medium of instruction. Singapore uses English, whereas Malaysia uses Bahasa Malaysia, the national language, as the medium of instruction in public tertiary institutions, while English is used in the private institutions. Language policies have definitely affected the development of the academic profession in the two countries.

In Singapore, student admissions to institutions of higher learning are merit based. However, in Malaysia, admissions to public institutions are based on an ethnic quota system. In attempting to promote social equity, the Malaysian government provides greater educational and occupational opportunities to Bumiputras[1] in the higher education system. Preferences are also given to Bumiputras, with respect to recruitment into the academic profession, especially at public higher education institutions.

In both of these countries, the academic profession has had to adapt to the rising student numbers, financial constraints, and the changing role of the state vis à vis higher education. The shift from elite to mass higher education, from higher education as a public service to higher education as a commodity, and from the role of the state as provider to that of regulator has vast implications for the academic profession. This chapter presents the argument that the strong intervention of the state combined with the influence of a market ideology have resulted in a hybrid bureaucratic and corporate academic culture in Malaysia and Singapore.

MASSIFICATION OF HIGHER EDUCATION

As in many other countries, higher education in Singapore and Malaysia has undergone rapid expansion due to ever-increasing social demand, which was brought about in part by the democratization of secondary education and the growing economic affluence of these two societies. In 2001 there were 11 public universities, 10 private universities, 6 polytechnics, 27 teacher training colleges, and about 600 private colleges in Malaysia. The total number of students enrolled at the tertiary level, in both the public and private sector, has doubled from about 230,000 in 1990 to about 550,000 in 1999. The enrollment rate within the 19-to-24-year age cohort has risen from 2.9 to 8.2 percent since 1990 (Ministry of Education, Malaysia, 1989, 1999). This rapid expansion has resulted in an acute shortage of academics in both the public and private sectors. Many young academics with only master's degrees were recruited by public universities and sent for further education and training as they began their careers in the universities. The situation is worse at private colleges, where one can find many first-degree holders teaching in first-degree programs.

In the case of Singapore, the expansion of higher education occurred in a more cautious and calculated manner. The government is committed to maintaining the overall efficiency of the system so as to avoid overproduction in certain fields, which would lead to graduate unemployment. In 1990, there were a total of about 63,500 students enrolled at the tertiary level (Ministry of Education, Singapore, 1999), and the enrollment had increased to about 127,500 in 2000 (Singapore Department of Statistics, 2000). These figures do not include students studying at privately run institutions. In 1992, 16 percent and 29 percent of the relevant age cohort were enrolled at local universities and polytechnics, respectively (Tan, 1997). Singapore has also been moving toward a mass system of higher education—targeting an enrollment rate of 60 percent at the tertiary level (20 percent in the universities and 40 percent in the polytechnics) in the year 2000 (Selvaratnam, 1994). The latest statistics show that this target has not been met yet, for in 2001 only 55 percent of the population had either secondary or higher education (Singapore Department of Statistics, 2001). The higher education system in Singapore is made up of two public universities, one private university, four polytechnics, one institute of education, and ten institutes of technical education.

Bureaucratization of Educational Institutions

As they have grown, universities have become more bureaucratic and complex. The dramatic growth in student numbers, academic staff, physical facilities, and support services requires the standardization of day-to-day operations. For example, the National University of Singapore has approximately 27,000 students, 2,500 academic staff, and 3,000 administrative staff (NUS, 1999). Thus, administrative staff outnumber academic staff. As universities expand, the direct power of academics over the institutional structures of governance has been limited by a new layer of professional bureaucrats who have significant power in the day-to-day administration of the university (Altbach, 1991).

As all public institutions are funded by the government, they are also run like government departments. Tertiary institutions in Singapore and Malaysia come under the direct purview of the Ministry of Education. Following the British tradition, the governance of the universities used to consist of the council, senate, central administration, faculties, departments, and specialized institutes and centers. The council oversaw the financial and administrative management, while the senate handled all academic issues of the universities. However, the recent reforms have changed this governance structure. Together with these changes in institutional governance, governments are demanding greater accountability regarding the expenditure of public funds by universities. Academics must submit to more fiscal control, increase their productivity, and be subject to more rules and regulations as well as rigorous assessment procedures. As a consequence, the academic culture is losing its collegiality and becoming more bureaucratic and hierarchical, with a concentration of power at the top.

THE PRIVATIZATION OF HIGHER EDUCATION

In the past, both the Malaysian and Singapore governments have been the main providers of higher education, providing complete funding to all public institutions of higher learning through budget allocations as well as block grants for development and capital expenditures. To expand access to higher education, the governments have kept tuition fees low by heavily subsidizing all public higher education institutions. Singapore, which is rich with resources, has had no difficulty in funding the expansion. However, in Malaysia, which has faced tight budgetary constraints, the state has had to relinquish its role as the main provider and to encourage the private sector to play an active role in the provision of higher education.

Private higher education has expanded tremendously in Malaysia in the last decade. The number of private institutions of higher learning has increased fourfold, from 156 institutions in 1992 to about 600 in 2000. In 1995 there was not one private university, but by 2001 there were ten. Most of the private institutions are profit-oriented enterprises although some are nonprofit. The for-profit institutions were set up by individual proprietors, private companies or consortia of companies, and publicly listed corporations or government corporations. The nonprofit private institutions were set up by foundations, charitable organizations, and through community support (Lee, 1999). These private institutions offer a wide range of programs in various fields, from the preuniversity to postgraduate level. The number of students enrolled at these institutions rose from about 35,600 in 1990 to about 250,000 in 1999, which accounted for 45 percent of the total number of tertiary-level students.

The liberalization of government policies toward private higher education is due to the lack of enough places at public institutions of higher learning to meet the increasing demand. This problem has been further exacerbated by the ethnic quota system for admissions to public higher education institutions, the high cost of overseas education, and the devaluation of the Malaysian ringgit during the recent Asian economic crisis (Lee, 1999). While the opening up of the private sector means more job opportunities for academics, it may also result in the deterioration of working conditions, especially for academics at small and poorly funded private colleges.

The private sector plays a less significant role in Singapore. A small number of private higher education institutions exist—with the largest being the Singapore Institute of Management, which was established in 1964. The institute enrolled about 6,000 students in 1992 (SIM, 1992), and in 1994 it was upgraded to a full-fledged university with a one-time financial grant and a donation of some land from the government (Selvaratnam, 1994). Two other private, nonprofit art and design colleges offer a variety of art and music programs. Each of these colleges enrolled fewer than 1,000 full-time students (Tan, 1997).

Through privatization, higher education becomes commercialized and commodified, which changes the working environment of the academic profession. Universities are now being viewed as business enterprises, and academics are encouraged to view themselves as "entrepreneurs." As observed by Buchbinder (1993), universities are beginning to operate like business organizations. Instead of producing and transmitting knowledge as a social good, universities emphasize the production of knowledge as a marketable good and a salable commodity. The market determines the courses taught, the research funded, the student markets served, and the enrollment policies adopted. In this corporate culture, the research output of university professors may not be evaluated by the traditional criterion of "is it valid?" but by commercial criteria such as "of what use is it?" or "is it salable?" (Currie, 1998).

CORPORATIZATION OF PUBLIC UNIVERSITIES

The penetration of the corporate culture into academia is not restricted to private educational institutions but extends to public universities. In 1995, the 1971 Universities and University Colleges Act was amended to lay the framework for corporatizing all public universities in Malaysia, and by 1998, five of the older public universities had been corporatized (Lee, 1999). Through corporatization, public universities are to be freed from the shackles of government bureaucratic supervision and are supposed to operate like business corporations. Corporatized universities are empowered to borrow money, enter into business ventures, raise endowments, and set up companies, among other activities. While the Malaysian government would continue to own most of the public universities' existing assets and provide development funds for new programs and expensive capital goods, the corporatized universities would have to raise about 30 percent of their operating costs. However, because of the recent Asian financial crisis, the Malaysian government has shelved its plan to reduce funding to public universities. Therefore, public universities were "corporatized by governance" only, not financially. Public universities are thus required to adopt the trappings of corporate culture—such as mission statements, strategic plans, total quality management, benchmarking, and staff development. This major reform is already having repercussions on the working conditions of the academic profession in Malaysia. Originally, the salaries of academic staff at corporatized universities were supposed to increase by 17.5 percent, but this salary increase has also been postponed indefinitely due to the financial crisis.

The same trend can be observed in Singapore. In June 2000, the key recommendations of a government committee that had been comprised to review the governance and funding of the two public universities were accepted by the government (Ministry of Education, Singapore, 2000). The committee recommended that the two public universities shed their civil service salary structures and pay according to performance and market factors. This change was deemed necessary to foster an entrepreneurial climate in the universities. Academic staff would be rewarded according to their performance, and the competitive pay scheme would attract "global talent" to work in Singapore's universities and help to propel them into world-class status. According to the vice-chancellor of the National University of Singapore (NUS), the performance-based, market-driven pay system for lecturers would help the university do away with the bureaucratic mindset and nurture the entrepreneurial spirit (Ideas and Dreams, 2000).

INTERNATIONALIZATION OF HIGHER EDUCATION

Higher education is fast becoming internationalized throughout the world, and Malaysia and Singapore are no exceptions. Internationalization is reflected in the flow of knowledge, scholars, and students across national borders. With the advances in information and communications technologies, many educational programs are being delivered in nontraditional modes like distance learning, off-shore programs, and e-learning. If in the past we talked about the "traveling salesman," today we have the "jet-setting professor."

In Malaysia, the influx of transnational education programs has emerged along with the development of private higher education. Many foreign universities from Australia, the United Kingdom, Canada, and the United States have set up institutional arrangements with local private universities and colleges to offer their educational programs to students on Malaysian soil. Some of them—like Monash University, Curtin University, and Nottingham University—have even set up branch campuses in Malaysia. These institutional arrangements can take various forms—such as twinning programs, credit-transfers, external degrees, distance learning, and joint programs (Lee, 1999). The availability of transnational educational programs has attracted many foreign students in the region to Malaysia. The number of foreign students has increased steadily from 5,635 in 1996 to 23,000 in 2000.

As for Singapore, universities there have set their sights on competing on the world stage. One strategy is to attract top-notch academic talent by offering competitive pay packages, in the hope of achieving world fame through research. Another strategy is to establish a "science hub" in Singapore that would include branch campuses of prestigious universities such as Johns Hopkins University and the French business school, INSEAD. A third strategy is to increase the number and percentage of foreign students by launching aggressive recruitment drives in the region. Both Singapore and Malaysia aim to be "centres of educational excellence" in the region.

THE ACADEMIC PROFESSION

The academic career structure in Malaysia and Singapore is a hybrid of the British and American systems. The faculty are ranked as professors, associate professors, lecturers, and tutors in Malaysia, with an additional two ranks—senior lecturers and assistant professors—in Singapore. For a long while, Singapore followed the British system, but it recently adopted the American ranking system to ensure that faculty nomenclature does not impede staff recruitment from outside the country. The total number of academics in the public universities in Malaysia was 10,920 in 1999, of which 5.6 percent were professors, 18 percent associate professors, and the rest lecturers. The number of academic staff members in all publicly funded polytechnics is

2,132, and the number in teacher training colleges is 2,698 (Ministry of Education, Malaysia, 1999). Figures for the private educational institutes are not available. In the case of Singapore, there are a total of 2,826 academics at universities, 3,931 at polytechnics, and 561 at the National Institute of Education (Singapore Department of Statistics, 2000).

The number of female academics varies across institutions and departments, but on the whole their proportion is much lower than that of male academics. Using the statistics from the NUS and Universiti Sains Malaysia (USM) as the basis for comparison, it is interesting to note that the proportion of female academics in each of these universities does not differ very much (NUS, 1999; Wazir, 1999). The proportion of female academics at the NUS is 31 percent, slightly higher than that at the USM, which has 25 percent. Only 7 percent of the professors at the NUS are women and at the USM, 6.6 percent. Among associate professors, 13.3 percent are women at the NUS and 15.2 percent at the USM. At the NUS 20 percent of the senior lecturers and 25.7 percent of the assistant professors are women, whereas at the USM 31.3 percent of the lecturers are women. However, these aggregate figures camouflage the fact that most of the women academics at the USM are found in the humanities, social sciences, education, and mass communications. As expected, there are very few women academics in the traditionally male-dominated disciplines like engineering, the natural sciences, and applied industrial technologies.

Although there is no salary differential between male and female academics, there does seem to be a "glass ceiling" blocking women's career advancement in the higher education sector. This brief analysis shows that the academic profession is still very much male dominated, especially if one takes into account the obvious absence of women in top-level positions at the university—such as vice-chancellors, deputy vice-chancellors, and rectors. The underrepresentation of women in the higher ranks of the academic profession can be attributed to sociocultural factors as well as organizational and structural factors (Luke, 1998). Like many other working women, female academics experience a "double-shift day" by pursuing a career and at the same time taking care of a family. Married women are at a disadvantage when compared to their male colleagues because of the domestic demands made on them as wives and mothers. This is particularly the case for married Malay women because Malay families tend to have more children when compared with other ethnic groups. Thus female academics would have less time and energy for research and publications, causing them to lose out to their male colleagues with regard to promotion up the academic ladder. Women academics are often not considered for promotion to high administrative posts, and even when offered positions they sometimes turn down the promotion because of their heavy family responsibilities. To illustrate the point, there is only one female dean at the USM, and she is the dean of the School of Educational Studies.

Academics at public educational institutions are very much part of the civil service although there have been recent moves to free them from the civil service salary structures. The civil service does offer tenure to academics. Once an academic is confirmed in a position, he or she is absorbed into permanent service until retirement age, at which time the retiree receives a government pension. An academic is entitled to a standard annual increment according to the relevant salary scale, and promotion is usually based on seniority. So far, academics are not unionized, but there have been attempts to form unions among academic staff in one or two of the private colleges in Malaysia. However, academics in Malaysia do have quite active academic staff associations, which are nationally as well as institutionally based.

The nationally based academic association known as Academic Movement Malaysia (Pergerakan Tenaga Akademik Malaysia) was formed in 1995, and it organizes conferences and issues publications on topics relating to the academic profession. As for the institutionally based associations, some are more active than others. For example, the Academic and Administrative Staff Association at the Universiti Sains Malaysia took the university to court for wrongful dismissal of two of its members and won the case (PKAPUSM, 1986). On a different front, the Academic Staff Association at the Universiti Malaysia negotiated with the government for a 17.5 percent pay raise across the board for all academic staff when the university was corporatized in January 1998. But because of the economic downturn, the Malaysian government has not upheld the collective agreement—which illustrates how little power academic staff associations have to improve the working conditions of the academic profession. Singapore has only institutionally based academic staff associations, which are very much subdued in their activities (Gopinathan, 1989). In fact, most of these academic staff associations do not attract many members, for academics in these two countries tend to shy away from activities that may be perceived as anti-establishment. They would rather be members of their own professional associations, both at the local and international level, which would help them to network as well as develop professionally. Local professional bodies like the Bar Council, Malaysian Medical Association, Malaysian Institute of Engineers, and others do get strong support from their members, whereas local discipline-based associations like the Malaysian Social Sciences Association, Malaysian Educational Research Association, and others are generally quite weak. On the whole, academics in Malaysia and Singapore, as a group of intellectuals, do not exert much power and influence in the governance of their institutions or in the public sphere.

GOVERNANCE AND POWER

The governance of all Malaysian public universities and colleges comes under a common legislative framework known as the 1971 Universities and University Colleges Act (UCCA). Under this act, the government has full authority over student enrollments, staff appointments, educational programs, and financing. The appointment of vice-chancellors and deputy vice-chancellors is placed in the hands of the minister of education. The election of deans was abolished, with the vice-chancellor given the prerogative to appoint them instead. The 1975 amendment of the UCCA forbids students or academics from becoming involved in any political activities or affiliated with any political party or trade union. The Malaysian government has used legislation to block both the dons and students from participation in shaping public discourse and national debates (Lee, 1997b). For instance, academics were discouraged from expressing their views publicly in the mass media about the economic crisis of 1997, the "haze incident" in 1998 (air pollution caused by fires in Indonesia), and the Nipah-Hendra virus epidemic in 1999.

The UCCA was further amended in 1995, but this did not remove the restrictions imposed on academics. Under the new guidelines, the university court was abolished, the university council replaced with a board of directors, the size of the senate reduced from about 300 to about 40, and universities allowed to engage in market-related activities. The reduction in the size of the senate may be viewed as a further erosion of academics' power in the governance of the university. Traditionally, the senate was comprised of vice-chancellors, deputy vice-chancellors, deans, faculty representatives, and all professors and was the place where policies concerning academic matters were thrashed out. A trimmer senate means less consultation and feedback on university policies for academics. A common anxiety among academics is the fear that intellectual pursuits might be sidelined by economic imperatives. Academics in Malaysia are subject to government control and pressured by the corporate culture that is rapidly emerging in the public universities.

In a corporatized public university, how much influence academics have on the decision-making process depends on how willing the vice-chancellor is to involve the academic staff. The vice-chancellor has the power to restrict the decision-making process to a small in-group, by appointing like-minded academics to the senate. With the institutionalization of corporate management practices at the universities, the vice-chancellor becomes more like a chief executive officer, making top-down decisions in response to the external environment. The move toward corporatization has eroded the traditional collegial mode of decision making in which decisions were made by academics on a democratic basis through councils and committees. The current thinking in the corporatized universities is that there is no time for lengthy deliberations and large committees because decisions have to be made fast so as not to miss out on all the opportunities available in a competitive market.

Academics working at private institutions of higher learning have even less say in the governance and management of their institutions. Private institutions are often governed by a board of governors or directors, usually comprised of representatives from the stakeholders. Although at some institutions the governance body is separate from the management team, at others the board and the management team consist of the same group of people. Private educational institutions owned by consortia of companies, publicly listed companies, and government corporations usually have a governing board that formulates policies and a management team headed by a chief executive officer who implements the policies on an operational basis. Academics who are not part of the management team do not have much say in the running of the institution. Their main role is to teach and make sure that students pass their examinations. At some private colleges, students are treated even better than their lecturers. The students are treated as highly valued "customers" whose interests and satisfaction must be attended to because they pay for the services offered by the colleges. For example, the management at one particular private college in Kuala Lumpur assigned the limited number of parking spaces on campus to students rather than to staff.

In Singapore, the government maintains strong control over the policy direction of all public tertiary institutions. The government, through its interventionist policy, has constantly directed tertiary institutions to respond and adapt to major societal changes and needs (Selvaratnam, 1994). All these institutions are governed by managerial rather than collegial or academic principles. Therefore, the majority of academics feel that they have very little say in decision making and hardly any influence in setting academic policy. The bureaucratic practice of making top-down decisions even on academic matters has deprived the academic staff of their right to be consulted.

However, university governance was put under review in 2000, and changes are taking place at two levels: first, the external level involving the interface between the Ministry of Education and the universities and, second, the internal level involving the organizational administrative structure within each university (Ministry of Education, Singapore, 2000). At the external level, although more academic and operational autonomy is given to the universities, they still have to comply with broad government policy guidelines on financial and personnel management. Internally within each university, more authority would be devolved to the dean's level or even below, to allow greater agility in responding to challenges at the faculty level. The move toward greater devolution of authority and financial resources to deans does give them some say in rewarding the staff financially, although major decisions such as tenure and promotions still lie at the university level.

FACULTY APPOINTMENTS

As mentioned earlier, the academic career structure at all public institutions in Malaysia lies within the civil service. After a person has been appointed to a faculty position, he or she has to go through a three-year probation period before being granted permanent tenure. Nearly all local appointees are confirmed after the probation period, and after confirmation they have full tenure rights until retirement. After ten years of service, academics are put on the pension scheme. The mandatory retirement age is 55 years, which is very young when compared to other countries. After much pressure from the civil servants union, the government announced in March 2001 that the retirement age would be raised to 56 years. The minimum academic qualification for a faculty appointment is a master's degree, but most faculty members at public universities hold a Ph.D. degree.

In the case of private institutions, there is a very acute shortage of well-qualified academics, and most of their staff do not hold any postgraduate degrees. Lecturers at private colleges tend to hop from one college to another in search of better prospects. To prevent them from job-hopping, tenure is given to a full-time lecturer after a three- to six-month probation, and there is no mandatory retirement age. At public universities, there is very little institutional mobility among academics except after retirement, when quite a number of them do

take up appointments at private colleges. The salaries of lecturers
teaching in the private sector vary from institution to institution and
are generally not very competitive, except in the case of those who are
teaching at private universities and at some of the more established
and larger colleges.

The private colleges also hire a substantial number of part-time
lecturers—especially the smaller colleges—partly to reduce costs and
partly due to the shortage of academics. Most part-time lecturers are
people who hold full-time jobs elsewhere. A part-time lecturer may
be an academic working full time at an educational institution who
has decided to take on a part-time job at another institution for extra
income, or he or she may be a professional working full time in industry and teaching on a part-time basis at a private college. There are
some lecturers, though not many, who earn their living by holding a
few part-time teaching jobs at several colleges, and such lecturers are
usually very poorly qualified academics who cannot find full-time
teaching jobs at any of the colleges.

Part-time lecturers are paid on an hourly basis, at rates that vary
across institutions and programs. The rate can be as high as U.S.$50
an hour in areas of specialization with high market value like information communications technology, law, and medicine. Part-time
lecturers are usually employed in educational programs with low and
unpredictable enrollments—such as those that are offered at poorly
funded colleges. The employment of part-timers does vary a lot,
ranging from 10 to 20 percent of the academic staff at some of the
more established colleges to as high as 80 to 90 percent at many of
the newer and smaller colleges.

Young lecturers teaching at private colleges usually do not intend
to make a career in academia. They take on these teaching positions
as stepping-stones to more lucrative careers in the private commercial
and industrial sectors. Therefore, it is much more difficult for private
colleges to attract teaching staff during economic boom times than in
economic hard times. Private colleges also seek to attract those who
are educated overseas and are more fluent in English than in the
national language. In fact, a large proportion of the academic staff
employed at private colleges are non-Bumiputras who obtained their
advanced education overseas because they were unable to gain admission to public universities in their own country due to ethnic quotas.
The working conditions at private colleges are not comparable to
those at public educational institutions: workloads are much heavier
and the facilities are far from adequate. At many private colleges, lecturers lack their own offices but have to sit together in a big staff

room just like teachers at a secondary school. Most colleges do not provide computer facilities for their teaching staff, and the libraries are inadequate to meet the lecturers' needs for teaching, not to mention for research purposes.

In Singapore, academics are still in the civil service—that is, until they are removed from it as recommended in the 2000 Report on Public University Governance and Funding in Singapore (Ministry of Education, Singapore, 2000). Unlike in Malaysia, there exists a strict tenure policy through which only 40 percent of the academic staff is tenured (Selvaratnam, 1994). To be tenured, a local staff member has to fulfill at least two three-year contracts; show teaching and research capabilities; and publish, particularly in international journals. As one senior academic put it, "Tenure is going to go, and merit pay is in. The university has the option to refuse tenure or decide when to give tenure." In view of this new development, many junior academics do feel pressure to do research and publish despite their heavy teaching duties. It is a common practice in Singapore for the universities to assign more teaching hours to their junior faculty members than to senior members. The retirement age in Singapore is later, at 60 years, but academics may choose to continue working until age 65.

COMPENSATION AND PROMOTIONS

Academics enjoy a relatively high social status in Malaysian society, but this is not reflected in the salaries they earn. Faculty members are paid and treated like any other civil servants. The main determinant of salary level is seniority and length of service at a particular rank, but faculty do get many compensatory benefits like housing and car loans, medical benefits, and sabbatical leave. With these benefits, they are able to lead a middle-class lifestyle. Unlike their counterparts in Indonesia and the Philippines, Malaysian academics seldom have to take on a second job to earn extra income to support their families. Most of them are able to maintain a comfortable standard of living, which includes owning a house and a car, sending their children to college, and even taking occasional holiday trips overseas.

Currently, all faculty members employed at public higher education institutions are covered under the New Remuneration Scheme, which offers a better deal to senior faculty but is not attractive to junior faculty. The salary scale ranges from U.S.$743 to U.S.$1,403 a month, for a lecturer; from U.S.$1,307 to U.S.$1,798, for an associate professor; and from U.S.$2,560 to U.S.$3,979, for a professor. Senior professors earn as much as five times as much as the most junior incumbents. Thus, public universities are encountering problems recruiting new lecturers despite the fact that these institutions have their own academic training programs for Bumiputras. As mentioned, these programs provide special opportunities to selected Bumiputra graduates to undertake postgraduate studies leading to higher degrees under the sponsorship of the universities.

Not only do the public educational institutions have to compete with the private sector for new graduates, they are also losing many of their experienced academic staff to private firms that offer much higher salaries. The exodus of academics to the commercial and industrial sectors has been gathering momentum since the early 1990s. For instance, the attrition rate of faculty members at the USM increased from just 7 percent in 1990 to 19 percent in 1995 and 27 percent in 2000. Academics leave the public universities partly for greener pastures in the private sector and also because of the overtly bureaucratic culture and ethnic policies at universities.

One of the consequences of the ethnic policies is the lack of promotion opportunities for non-Bumiputra academics at public universities. Except for a few token deans, there are hardly any non-Bumiputras in senior administrative positions. Although all qualified faculty members are entitled to apply for promotion to associate professor or professor, very often actual promotions are decided on the basis of the ethnic quota. The process of promotions is neither transparent nor very democratic. An applicant first submits his or her curriculum vitae. After the application is reviewed by a university committee and the applicant is interviewed by a panel including the dean, the short-listed candidate has his or her curriculum vitae assessed by two external evaluators, whose evaluation may or may not carry weight depending on whether or not the university authorities want to promote the candidate. The ethnic policies at Malaysian universities have caused much frustration and many grievances among non-Bumiputra academics, resulting in quite a number of them migrating to work in countries like Singapore, Hong Kong, and Australia.

Academic salaries in Singapore are high, and faculty are entitled to many fringe benefits. The recently revised salary scales are internationally competitive and rank among the highest in the region. The universities have a pay scale of about U.S.$2,750 to U.S.$4,650 a month for assistant professors, about U.S.$3,900 to U.S.$8,200 for associate professors, and about U.S.$6,900 to U.S.$9,750 for professors (Ministry of Education, Singapore, 2000). Besides the monthly pay, staff members are allowed to undertake outside consultation, and under the new salary scheme they can keep all their consultancy fees. The universities, however, do impose a time limit on consulting work. Although academics earn very good wages, the cost of living in Singapore is also very high.

Promotions at Singapore universities are based on a point system that includes teaching, research, service, publications, and administrative responsibilities, as well as peer review. An assistant professor applying for promotion to associate professor has to be recommended by his or her head of department and examined by departmental and faculty review committees before being called for an interview before the university review committee. The same procedure applies for promotion to full professor, except that external assessors would review the candidate and the vice-chancellor (rather than the deputy vice-chancellor) would chair the university review committee. Such promotion exercises are carried out on an annual basis. With the implementation of the new remuneration system as recommended by the 2000 Review of Public University Governance and Funding in Singapore, the criteria for promotion will be based on performance rather than on seniority (Ministry of Education, Singapore, 2000).

In both Malaysia and Singapore salaries and raises for academic staff at public universities are merit based, but the system operates quite differently in the two countries. The Malaysian civil service salary structure includes three subscales for each salary category—P1, P2, and P3. The salary points in P2 are 4.5 percent better than those in P1; similarly, those in P3 are 4.5 percent better than those in P2. The dean has the authority to recommend only 5 percent of the academic staff annually to move from the P1 subscale to either P2 or P3, based on merit. However, since the amount of money involved is very small, as is the number of persons that the dean can recommend to benefit from these increments, most academics do not take this incentive very seriously. In the case of Singapore, the merit-based salary scheme does make a lot of difference in the amount of money drawn by a high performer as compared to a low performer. The new remuneration system consists of a basic component and other variable components reflecting differences in performance, responsibilities, and market value. There is no automatic annual raise because pay increases within a salary range are to be based on performance and the recommendation of the immediate superior. To ensure that this merit-based and flexible salary scheme works fairly and efficiently, a rigorous appraisal system for the academic staff is to be institutionalized at each university. This salary reform has created a lot of anxiety among average and low-performing academics, but to the high flyers it is a long overdue change.

RECRUITMENT OF ACADEMICS

Academic staff recruitment strategies differ greatly between Singapore and Malaysia—one is outward looking and the other is inward looking. While Singapore has positioned itself to recruit from the international arena, Malaysia continues to recruit more Bumiputras into academia, in line with its affirmative action policies. In the past, Malaysian universities sent many of their newly recruited staff away for further studies and professional training, but because of the 1997 economic crisis and budget cuts most junior staff now do their postgraduate studies at local universities. As a result, there are too many cases in which a faculty member has obtained all his or her degrees from one particular university, while working under the same professor. Furthermore, because Bahasa Malaysia is the medium of instruction, finding external examiners outside the country to examine Ph.D. theses is extremely difficult. These constraints lead to inbreeding in academia due to the lack of cross-fertilization of ideas.

On the other hand, Singapore realized that its locally available pool of academic talent was inadequate to meet the demand for a wide range of specializations. Therefore, the country embarked on a policy of robust international recruitment so as to ensure that only staff with high professional and academic standards were recruited. Singapore competes for talent on the global market. Expatriates comprise as much as 20 percent of the academic staff at universities and 12 percent of the staff at the polytechnics (Selvaratnam, 1994). The majority are from the United States, the United Kingdom, Australia, New Zealand, India, China, Hong Kong, and Taiwan. Singapore's tertiary institutions expect the expatriate academics to provide new ideas and new areas of research, and to introduce sound teaching approaches. However, they are not given tenure; instead, they are employed on a contractual basis with many fringe benefits like travel, a settling-in-allowance, housing, gratuity, home leave, educational allowances, medical and dental coverage, and car loans. In spite of these attractive packages, expatriates do not stay long in Singapore because they feel insecure and alienated, especially when they are being employed on short-term contracts. They have little or no opportunity for a career path and growth and are thus not committed to the mission of their institutions. They tend to devote their time, at the expense of teaching, to discipline-oriented research in order to enhance their academic standing internationally so that they can seek employment with tenure elsewhere (Selvaratnam, 1994).

The presence of expatriates in the academic profession in such large numbers in Singapore is quite a unique situation and one that can create tensions and contradictions. The tension between expatriate and local academics can be traced back to the late 1960s, when the academic staff at the NUS formed two separate academic staff associations—one for the expatriates and the other for the locals, because of differences in their interests and culture. The expatriates were more concerned with their terms and conditions of employment, and the locals felt that the expatriates were not able to understand the aspirations of the nation and uninterested in bettering the terms and conditions of service of local staff. It should be noted that in the early years after independence most of the top administrative positions in the university were held by expatriates (Gopinathan, 1989). This tension has continued until today although it has taken on different forms. Singaporeans are generally very competitive and in a big hurry to catch up with the advanced industrial countries and are thus very receptive to change and to new ideas, which they expect the expatriates to provide. As an assistant professor at the NUS observed, "We are in a great flux with rapid changes just like an earthquake zone. We are always trying to improve and to get the best from the West. Some of us still exhibit the colonial mentality by looking up to the Whites for academic leadership."

However, there is some resentment among local academics that expatriates get more perks and fringe benefits than do the locals. For example, a senior professor remarked, "Why should we pay the expatriates more [when] we can attract the best talents from the world on our own terms?" Locals also resent the condescending attitude of some expatriates. It was interesting to hear an expatriate complain, "They are not making full use of my expertise here." Despite all these undercurrents, both locals and expatriates do work together relatively harmoniously, which results in fruitful collaboration in research or publications.

In the case of Malaysia, there are very few expatriates working at local educational institutions, except at foreign branch campuses and some private institutions. Most of the expatriate academic staff left the country when Malayanization took place in the 1960s. Later, the implementation of the New Economic Policy in 1970 signified the beginning of "Bumiputra-ization" of the public universities. Academic staff recruitment was no longer based on merit but rather on ethnicity. Universities must recruit large numbers of Bumiputra staff to create the required ethnic balance, and many young Bumiputra staff members have been rapidly promoted to key administrative positions as heads of department and deans (Singh, 1989). Since 1974, the appointment of deans and heads of departments has been the prerogative of the vice-chancellors, who strive to assemble a smoothly functioning and cooperative team—in many instances, the appointments were not based on academic leadership skills (Ahmat, 1982).

The "Bumiputra-ization" policy has also greatly reduced the proportion of non-Bumiputra academics working at public tertiary institutions, thus resulting in an emerging ethnic divide between the public and private sector. As mentioned earlier, the majority of academics teaching in the private sector are non-Bumiputras, while most of those in the public sector are Bumiputras. This ethnic divide is further exacerbated by the fact that 90 percent of the students enrolled in the private sector are also non-Bumiputras, whereas the majority of students at public institutions are Bumiputras. This divide between public and private higher education is a concern, and the Malaysian government has implemented various means to address the problem, but so far with no success.

THE BUREAUCRATIC ACADEMIC CULTURE

From the 1960s until the early 1990s, public universities in Malaysia also became very bureaucratic and hierarchical in their day-to-day administration, and this bureaucratic culture has strongly influenced the daily lives of academics. Because of their low salaries, especially at the beginning of their service, university lecturers do their best to obtain an administrative position because the monetary allowance that comes with the position supplements their monthly income. To enhance their chances of being promoted, they follow every rule and regulation laid down by the administration and carry out their academic work like bureaucrats. Some of them even work strictly according to office hours, while others restrict their academic work solely to teaching, just like high school teachers. Many of them acquire the attitudes of a bureaucrat, refraining from expressing criticism and displaying blind loyalty to their department heads. Although there are no concrete data, some evidence does exist of lecturers being involved in side businesses (Salleh, 1991).

This academic bureaucratic culture, a result of too much direct government control over the universities, has degraded academia in Malaysia. The development of such a culture has also tightened the government's grip on the universities because academic bureaucrats turn to the government and to political leaders for recognition and rewards that have nothing to do with academic achievements.

Academics are no longer promoted on the basis of academic performance but rather according to nonacademic factors such as "favouritism, political patronage, administrative experience, and other kinds of cronyism" (Universiti, Para Akademik dan Masyarakat, 1985). As a consequence, many academics promoted to leadership positions are lacking in intellectual maturity and academic leadership qualities.

The situation is no better at private institutions, where hiring and firing lies in the hands of the owners or chief executive officers. Since many of these institutions are run on a commercial basis, very often lecturers are required to market their programs and recruit students, since if the programs are not economically viable, there is a likelihood they may very well lose their jobs. To a large extent, this bureaucratic academic culture has directly and indirectly affected the roles and functions of academics.

The typical functions of a full-time faculty member at any public university in Malaysia and Singapore include teaching, research, service, and, recently, consulting. Because of the rapid expansion of higher education, many lecturers have to carry heavy teaching loads, more so in Malaysia than in Singapore. Over the years, the ratio of staff to students has deteriorated from $1:20$ to as high as $1:40$, and the average number of teaching hours per week has increased from 9 to 12 hours per week in Malaysia. Of course, these figures do vary from department to department. In Singapore, the staff-to-student ratio has been kept quite low, at $1:10$, with much lower ratios in medicine $(1:6)$ and dentistry $(1:4)$. The polytechnics have a higher ratio of $1:14$. Average teaching loads range between 9 and 10 hours a week at the university and 20 hours per week at the polytechnics (Selvaratnam, 1994). However, the teaching loads at private colleges are much higher, ranging from 15 to 20 hours per week. Because of the increasing number of international students enrolled at these private colleges, the lecturers have to teach students from more culturally diverse backgrounds.

There is hardly any research or consultancy work among academics at private colleges and polytechnics. Academics at universities are expected to do research. However, commitment to research is quite erratic. It is possible to find academics who are very dedicated to research and others who barely do any research at all. The research productivity among Malaysian academics is generally low because the academic environment is not conducive to research. A study on the research productivity among academic staff at the Universiti Pertanian Malaysia shows that they spend about 30 percent of their time on research; active researchers produce on average one article per annum (Haris, 1985). The lack of motivation to do research can be attributed to several factors, such as lack of time due to heavy teaching load, lack of funding and equipment, and lack of academic leadership from senior faculty members. As observed by Singh (1989), the lack of academic leadership is partly due to the government policy of greater Bumiputra representation, especially in key administrative positions in the universities. Such a practice has not always resulted in the appointment of senior scholars and experienced scientists to key posts. For those who have moved early in their careers into administrative positions, research activity often receives low priority. Therefore, they are not able to provide strong research direction for their departments or schools because they themselves have never developed their own research capabilities. Furthermore, those who have the expertise but are outside the power structure were reluctant to provide this kind of leadership.

In the past three decades, public universities in Malaysia have been confronted with challenges that include rapidly increasing enrollments, the change in the medium of instruction from English to Bahasa Malaysia, and the Bumiputra-ization of the academic staff. The inability of Malaysian universities to produce cutting-edge research has led to the proliferation of scientific research institutes set up by the government to study specific problems relating to national development (Lee, 1997a). Examples of these research and development institutes are the Rubber Research Institutes, the Malaysian Agriculture Research and Development Institute, the Palm Oil Research Institute of Malaysia, the Malaysian Institute of Microelectronic Systems, and others. These research institutes are well funded by taxpayers' money, and through default, academics at the public universities are forced to compete with researchers at these research institutes for research funds. In cases in which academics are burdened with heavy teaching loads and a shortage of research funds, they have had to surrender their role as "knowledge producers" and be contented with simply disseminating knowledge.

The conversion of the medium of instruction from English to Bahasa Malaysia in the education system has brought with it a whole host of problems for the public universities in general and academics in particular. To facilitate knowledge dissemination, much effort has been made to adapt the Malay language for academic and scientific instruction. Many academics who are well versed in both English and Malay have spent much time and energy translating foreign works in their areas of specialization. The decline in the standard of English among Malaysian academics and postgraduate students has also affected the quality of research and development work in the universities. Because of their poor command of the English language, which is especially the case among people trained locally, academics are not able to keep up with the latest research and development in their fields of study at the international level.

The lack of motivation to do research in Malaysia can also be attributed to other factors, like the lack of recognition given to research and publications by universities. There is no formal regulation in the university specifying that research is compulsory for all and that appointments of tenured staff can be terminated if they have not engaged in research. The majority of academics believe that research and publications make little difference to their promotion. What matter are their political stance and connections (Osman, 1991). Another factor could be the lack of linkages between academics and local industry and between academics and the international intellectual community, which can make academics feel very isolated and lonely in their research efforts.

In Malaysia most research is commissioned by government agencies, statutory bodies, and private companies. This kind of research has advantages and disadvantages. On the one hand, the research may be relevant to the needs of the client or of the broader society. It also brings in financial rewards to both the researchers and universities concerned. In some cases, such arrangements open access to information that would otherwise be unavailable to academics. On the other hand, this kind of research tends to serve as a strong disincentive for academics to act as independent critics and commentators in society. Academics might be tempted to do little or no independent research, which means that many other important issues would remain unexamined. Furthermore, research quality would be more likely to fall because the results of commissioned research are usually not published and therefore would not undergo peer review. The result is that the research environment in Malaysia does not encourage cutting-edge research that might compete in the global arena.

In Singapore, the prospects for research are much brighter. There is no shortage of resources for research, especially in targeted areas like information communications technology, biotechnology, life sciences, microelectronics, robotics, and artificial intelligence (Tan, 1997). A study of the research environment in Singapore shows that the government is committed to providing financial support and incentives for research with economic potential, especially applied research in high-technology areas (Pang and Gopinathan, 1989). Unlike the situation in Malaysia, there are close ties between academics and industries. The study also shows that most academics are strongly research oriented and more cosmopolitan than local in their professional outlook. They are aware of the value and rewards obtained through publication in international journals. Overall, in both countries pressure to do consultancy work and to engage in revenue-generating activities has increased in recent years.

CORPORATE CULTURE

As mentioned earlier, the Singapore government wants its public universities to operate as entrepreneurial universities, with the vice-chancellors assuming the new role of CEOs of their organizations. To set the stage for this reform, steps were taken to change the governance, funding, and staff salaries at public universities (Ministry of Education, Singapore, 2000). The universities would be given a block grant every three years (instead of each year), and they would be allowed to decide on how to spend the money. They could retain surpluses but would have to make up any shortfall from their own funds. The compensation of academic staff would be removed from the civil service structure and be determined by market forces. To prevent a brain drain to more lucrative jobs in the private sector, an overall pay raise has been included, especially to the assistant professors, whose pay was set to go up in 2001 by as much as 20 percent, based on merit. The automatic, time-based increments would be abolished and academic staff would be paid and given increments based on performance. Only performance would decide promotions and increments. "Star" professors would be well rewarded with performance bonuses, fellowships, and one-off monetary awards. The performance-based, market-driven pay scheme would result in staff from different faculties receiving different pay packets and introduce greater variation of pay between top and average performers. This new pay scheme is aimed at attracting talent (staff and students) and improving education and research programs at the public universities.

Similarly, there have been attempts to remove the academic staff salary scheme from the civil service structure in Malaysia. When the corporatization of public universities was first conceptualized, each university was supposed to be given the autonomy to set its own salary scheme for its academic staff instead of following a common civil service scheme. With the removal of their civil servant status, academics are supposed to be able to negotiate their salaries and working conditions on an institutional basis. The rationale behind this move is to allow each institution to compete for talent in the academic labor market. However, this part of the reform did not go into effect for some reason, but there are other changes that have had a direct impact on the working environment of academics.

First, universities in Malaysia began to operate like business corporations and profit-making centers. Working in this corporate culture, academics are under increased pressure to obtain funding and revenue from external sources and to generate income for the universities. Among other things, they are required to do consultancy work, seek research grants, enroll full-fee-paying students, franchise their programs, and produce commercial patents from their research. They are allowed to engage in "market" activities in addition to their duties as "civil servants." For example, medical doctors working in the Faculty of Medicine at the Universiti Malaya are allowed to open private clinics in the UM Teaching Hospital outside their office hours. The emergence of a corporate culture in the universities is beginning to cause a cleavage between academics in the natural and applied sciences, who are constantly under pressure to engage in entrepreneurial activities, and those in the social sciences and humanities, who perceive the social value of their research as being undermined by the university authorities (Lee, 1999). Many Malaysian academics also fear that too much attention will be given to entrepreneurial activities at the expense of the "core business" of the university—teaching and research. The corporatization of public universities has brought about increased institutional autonomy but along with it more accountability on the part of academics.

ACADEMIC FREEDOM

Academics are professionals, but they also have to work in the bureaucratic environment of the university. In most societies, the academic profession traditionally values its autonomy and control over its own goals, the curriculum, degree requirements, and the ethos and orientation of the university (Altbach, 1991). The concept of autonomy relates to individual academic freedom and to the institutional autonomy of the university. Professorial and institutional autonomy is increasingly challenged by accountability, that is, "the requirement to demonstrate responsible actions to some external constituencies" (Berdahl, 1990, p. 171). To a large extent, the issue of autonomy and accountability is dependent on the relationship between academic institutions and the state, with the former insisting on more autonomy and the latter demanding more accountability.

As pointed out by Altbach (2000), academic freedom in Singapore and Malaysia is limited when compared to other countries. There are restrictions on what can be researched and what the academic community can express to the public. Issues such as ethnic conflict, local corruption, and other politically sensitive subjects have been banned as topics for academic research. Foreign researchers are required to obtain permission from the Economic Planning Unit before they may carry out empirical research in Malaysia, and educational researchers have to get approval from the Ministry of Education before they may enter schools to carry out research. There have been cases of censorship of research findings that were deemed to be politically sensitive by the powers that be. There is also a ruling that academics must seek permission from the vice-chancellor before expressing their views publicly. A recent example is the case involving an associate professor at the Universiti Malaya who was issued a "show cause" letter by the university over an Internet posting in support of the parents who objected to a move by the Ministry of Education to relocate a Chinese primary school (Professor's Fate, 2001). With such tight surveillance, many academics tend to self-censor in what they say and what they write.

In Malaysia, at the institutional level, academics have the power to assign textbooks, determine readings, and evaluate their students, but each new course needs to be approved by a studies committee set up by the university senate. In Singapore, within the broad policy guidelines, the universities are supposed to enjoy a considerable amount of freedom to select and admit students, design course content and delivery, and set their examination policy. However, university autonomy has long been eroded through a series of confrontations between the universities and the government—among them, the "Enright affair" in 1960, the issue of the "suitability certificate" in 1964, and the expulsion of several law undergraduates in 1966 (Selvaratnam, 1994). Academic freedom as defined in the Western sense is also very limited because the Singapore government plays a large role in determining the kind of programs and research carried out at universities (Gopinathan, 1989). Furthermore, the government is highly sensitive to criticism of its policies. In one instance, an expatriate professor was immediately deported for criticizing the government in the mass media.

However, academics teaching at private institutions in Malaysia have even less academic freedom because many of them are teaching in educational programs that they have no part in designing or even evaluating. In most offshore programs—such as twinning programs and credit-transfer programs—the curricula are developed and the examinations are set by the overseas partner institutions. In the new period of entrepreneurship, academics increasingly seek research funding from the private sector. This will have implications for academic freedom because the interests of these private firms may one day become dominant on campus.

The corporate culture in universities places a lot of emphasis on performance. Some university administrations have adopted such corporate management practices as implementing performance indicators, benchmarking, and management-by-objectives. Even academic activities have come under close scrutiny. For example, academic staff at all public universities in Malaysia have to work out "personal performance contracts" with their heads. They have to submit very detailed statements about the work they completed the previous year and a set of objectives for the following year. In the case of Singapore, academics are rewarded on the basis of performance and are subject to a rigorous appraisal system that includes features such as "use of multiple sources of information, goal-setting, self-appraisal, appraisal by supervisors, moderation by faculty-level committees, and an appeal process" (Ministry of Education, Singapore, 2000, p. 5). In short, there is not much institutional autonomy or academic freedom at public universities in Malaysia and Singapore for they are tightly controlled, supervised, and managed by the state.

FUTURE TRENDS

The expansion of higher education will continue until the targets of 60 percent and 40 percent of the age cohort in Singapore and Malaysia, respectively, are met. Public tertiary institutions will continue to be the dominant providers in Singapore, with private institutions establishing niches for themselves. In Malaysia, many of the older universities have already reached full capacity in student enrollments (about 20,000 at each university), but the newer universities will continue to expand. There will be a period of consolidation in the private sector, in which private educational institutions will compete, merge, or even die. Some private colleges will establish niche markets for themselves. Others will compete on similar turf, with their status and reputation ultimately determined by the quality of education they provide.

Academics, working in an expanding higher education system, will have no fear of not getting a tenured job. In Malaysia, the number of tenured positions in academia exceeds the supply of academics, especially for Bumiputras, at public tertiary institutions. With the ethnic quota policy firmly in place, an ethnic divide is clearly emerging between the public and private institutions. Not only are most students enrolled at private institutions non-Bumiputras, many of the private institutions are owned and staffed by non-Bumiputras. Whether this ethnic divide will deepen or lessen will depend to a great extent on the type of social engineering policies that the government puts in place with respect to higher education. Tenure in Malaysia is going to continue because of the acute shortage of academics in both the public and private sector. Although the mandatory retirement age for academics is 56 years at public institutions, academics will not lack for employment if they choose to continue to teach beyond the retirement age because they can always be reemployed by the government on a contract basis, or they can join private institutions. However, there will be a growth of part-time and non-tenure-track appointments at private institutions as they compete for human resources with the public institutions and private enterprises.

The scenario in Singapore differs slightly because the competition for academics is played out not in the local labor market but in the global market. The newly implemented competitive salary scheme will attract a large number of expatriates to work in Singapore and will hopefully help Singapore's universities to achieve world-class status. The performance-based salary scheme will definitely exert a great deal of pressure on the academic staff, who will hopefully respond by increasing their efficiency and productivity. The removal of civil service status from academics may help them to snap out of the complacent attitude that is typical of civil servants. The increased influx of foreign expertise into academia will continue to cause tension between the locals and expatriates as they compete for rewards, recognition, and promotions. Furthermore, tenure will be more difficult to come by as universities strive for institutional flexibility so as to keep pace with rapid social and economic changes.

In times to come, the civil service status of academic staff at corporatized public universities in Malaysia may be changed into a set of contractual relationships worked out bilaterally between each university and its academic staff members. This transformation may bring about more institutional mobility among academics as they go from university to university in search of better prospects and career development. At private institutions, the terms of appointment for academics will continue to be based on conditions settled between employers and employees on an individual basis unless academics are successful in forming unions to do collective bargaining.

Tertiary institutions in Singapore and Malaysia will continue to increase their intakes of foreign students as they compete to become the center of education in the Southeast Asia region. Some of the more enterprising educational institutions will venture into distance education through e-learning or offshore programs. In view of this developing trend, there is urgent pressure for professors to learn how to use computer and other educational technology to enhance their teaching. They will also need to learn how to cater to the diverse needs of their international students and be culturally sensitive both inside and outside the lecture hall.

As tertiary institutions compete for students and academic staff from within and outside the country, they will have to maintain, if not improve, the quality of their educational programs and research. They will also be competing among themselves at the international league tables, a ranking exercise that was carried out annually until 2001 by *Asiaweek* magazine (Asia's Best Universities, 2000). The establishment of quality assurance bodies, like the National Accreditation Board in Malaysia, means that academics will be more frequently subjected to quality audits and external evaluation of their performance.

The government will continue to have overall control and direct the development of higher education systems in both these countries. Therefore, the role of academics in the governance of universities will be minimal. The prevalence of the bureaucratic and corporate culture in universities may continue to erode the academic traditions of the professoriate even further. It is important to bear in mind that the contemporary university, in Malaysia and Singapore, is a Western transplant that does not have a long academic tradition. This nascent academic tradition is not very likely to withstand the onslaught of certain bureaucratic practices and market ideologies. Thus, the scholar is constantly being transformed into a bureaucrat or entrepreneur in Singapore and Malaysia.

NOTE

1. Bumiputra means "native of the soil." This term is used for the Malays and other indigenous tribes such as the Kadazans and Dayaks. The Bumiputras enjoy "special privileges," as enshrined in the Malaysian Constitution (see Article 152 of the 1957 Malaysian Constitution).

REFERENCES

Ahmat, Sharom. (1982). *Planning and management in the Science University of Malaysia*. Singapore: Maruzen Investment.
Altbach, P. G. (1991). The academic profession. In Altbach, P. G. (Ed.), *International higher education: An encyclopedia* (pp. 23–45). New York: Garland.

Altbach, P. G. (2000). The deterioration of the academic estate: International patterns of academic work. In P. G. Altbach (Ed.), *The changing academic workplace: Comparative perspectives* (pp. 1–23). Chestnut Hill, MA: Center for International Higher Education, Boston College.

Asia's best universities. (2000). Retrieved August 30, 2000 from <http://www.cnn.com/ASIANOW/asiaweek/features/uni-versities 2000/scitech/sci.overall.html>.

Berdahl, R. (1990). Academic freedom, autonomy and accountability in British universities. *Studies in Higher Education,* 15(2), 169–180.

Buchbinder, H. (1993). The market-oriented university and the changing role of knowledge. *Higher Education,* 26, 331–347.

Currie, J. (1998). Globalization practices and the professoriate in Anglo-Pacific and North American universities. *Comparative Education Review,* 42(1), 15–29.

Gopinathan, S. (1989). A university education in Singapore: The making of a national university. In P. G. Altbach and V. Selvaratnam (Eds.), *From dependence to autonomy* (pp. 207–224). Dordrecht, Netherlands: Kluwer Academic.

Haris, G. T. (1985). Constraints on research in Malaysian universities. *Southeast Asian Journal of Social Science,* 13(2), 80–92.

Ideas and dreams are the new buzzwords at NUS. (2000, August 24). *The Straits Times.* Retrieved June 9, 2000, from <http:// straitstimes.asial.com.sg/4/singapore/sin3_0824_prt.html>.

Lee, M. N. N. (1997a). Malaysia. In G. A. Postiglione and G. C. L. Mak, (Eds.), *Asian higher education: An international handbook and reference guide* (pp. 173–197). Westport, CT: Greenwood.

———. (1997b). Reforms in higher education in Malaysia. In K. Watson, C. Modgil, and J. Modgil (Eds.), *Reforms in higher education. Educational dilemmas: Debate and diversity,* Vol. 2 (pp. 195–205). London: Cassell.

———. (1999). Corporatization, privatization, and internationalization of higher education in Malaysia. In P. G. Altbach (Ed.), *Private Prometheus* (pp. 137–159). Westport, CT: Greenwood.

Luke, C. (1998). Cultural politics and women in Singapore higher education management. *Gender and Education,* 10(3), 245–263.

Ministry of Education, Malaysia. (1989). *Malaysian educational statistics, 1989.* Kuala Lumpur: Ministry of Education.

———. (1999). *Malaysian educational statistics, 1999.* Kuala Lumpur: Ministry of Education.

Ministry of Education, Singapore. (1999). *Education Statistics Digest, 1999.* Singapore: Ministry of Education.

———. (2000). *Fostering autonomy and accountability in universities: A review of public university governance and funding in Singapore.* Singapore: Ministry of Education.

National University of Singapore (NUS). (1999). *Facts and figures 1998/1999.* Singapore: NUS.

Osman, S. (1991). Politik akademia kini dan Mejelang Tahun 2000. In W. M. Wan Muda and H. Md. Jadi (Eds.), *Akademia* (pp. 34–43). Penang: PKAPUSM, Universiti Sains Malaysia.

Pang, E. F., and Gopinathan, S. (1989). Public policy, research environment, and higher education in Singapore. In P. G. Altbach, C. H. Davis, T. O. Eisemon, S. Gopinathan, H. S. Hsieh, S. Lee, E. F. Pang, and J. S. Singh, *Scientific development and higher education* (pp. 137–176). New York: Praeger.

Persatuan Kakitangan Akademik dan Pentadbiran Universiti Sains Association (PKAPUSM). (1986). *Krisis Universiti Sains Malaysia.* Pulau Pinang: Author.

Professor's fate lies in UM's hands, says Musa. (2001, March 21). *The Star,* p. 6.

Salleh, H. (1991). Perubahan dan Kontradisi Dalam Tradisi Akademia di Malaysia. W. M. Wan Muda and H. Md. Jadi (Eds.), *Akademia* (pp. 1–33). Penang: PKAPUSM, Universiti Sains Malaysia.

Selvaratnam, V. (1994). *Innovations in higher education: Singapore at the competitive edge.* Washington, DC: World Bank.

Singapore Department of Statistics. (2000). Facts and figures on education. Retrieved July 9, 2001, from <http://www.singstat.gov.sg/FACT/SIF/sif20.html.>

———. (2001). Key statistics—annual statistics on literacy and education. Retrieved July 25, 2002 from <http://www.singstat.gov.sg>.

Singapore Institute of Management (SIM). (1992). *Annual Report 1992.* Singapore: SIM.

Singh, J. S. (1989). Scientific personal research environment, and higher education in Malaysia. In P. G. Altbach, C. H. Davis, T. O. Eisemon, S. Gopinathan, H. S. Hsieh, S. Lee, E. F. Pang, and J. S. Singh, *Scientific development and higher education* (pp. 83–136). New York: Praeger.

Tan, J. E. T. (1997). Singapore. In G. A. Postiglione and G. C. L. Mak (Eds.), *Asian higher education: An international handbook and reference guide* (pp. 285–309). Westport, CT: Greenwood.

Universiti, Para Akademik dan Masyarakat. (1985). *Ilmu Masyarakat,* 10, 90–92.

Wazir J. K. (1999). *Women in higher education.* Penang: Universiti Sains Malaysia.

World Bank. (2001). *World development indicators.* Washington, DC: World Bank.

7

THE CHANGING ACADEMIC
WORKPLACE IN KOREA

Sungho H. Lee

At the start of a new millennium of global competition among institutions of higher education, Korean academics face formidable challenges. They must create bridges between the intellectualism of the ivory tower and the requirements of the world of work, bringing competitive new ideas and products to their institutions and to society. They must also help young people to get excited about academic inquiry, tackle growing social problems, and ensure that future generations of Koreans are capable of living and working in a multicultural, global society. In short, academics need to summon new strength and commit themselves to meeting the unprecedented challenges facing Korean higher education in the twenty-first century. The future of Korean higher education will remain intertwined with the role of academics in this rapidly changing academic workplace.

THE CHANGING ENVIRONMENT OF
THE ACADEMIC SYSTEM

Quantitative Growth

With the exception of some older colleges and universities, most modern institutions of higher education in Korea were established after World War II. When Korea was liberated from Japanese colonial rule in 1945, there were only 19 institutions of higher education, with 7,819 students and 753 faculty members. During the past half century, however, Korean higher education has undergone massive growth. The rate of growth appears to be leveling off now, as shown in table 7.1.

Table 7.1 Higher education institutions, students, and faculty, 2000

Type of institution		Institutions	Students	Full-time faculty
Junior colleges	public	16	37,331	740
and vocational	private	142	875,942	10,967
colleges (2–3 year)	total	158	913,273	11,707
Colleges	public	26	372,078	11,359
and universities	private	135	1,293,320	30,584
(4 year)	total	161	1,665,398	41,943
Industrial	public	8	81,186	1,253
universities	private	11	89,436	1,137
(4 year)	total	19	170,622	2,390
Teachers	public	11	20,907	698
universities	private	0	0	0
(4 year)	total	11	20,907	698
Air and	public	1	360,051	113
correspondence	private	0	0	0
universities	total	1	360,051	113
(4 year)				
Miscellaneous	public	0	0	0
institutions	private	5	3,861	52
(2–4 year)	total	5	3,861	52
Graduate	public	0(150)[a]	71,498	0
schools	private	17(662)[a]	157,939	179[b]
	total	17(812)[a]	229,437	179[b]
All	public	62	943,051	14,163
types	private	310	2,420,498	42,919
	total	372	3,363,549	57,082

Notes: Data from Korean Education Development Institute, 2000; Ministry of Education, 2000.
[a] The numbers in parentheses refer to graduate schools established in conjunction with four-year universities.
[b] The number of faculty members in graduate schools includes only those at the 17 independent graduate schools that have no undergraduate programs.

The total number of higher education institutions had increased to 372 by 2000: 158 two- or three-year junior vocational colleges; 192 four-year colleges and universities; 829 graduate schools, including 17 independent graduate schools that have no undergraduate programs; and 5 miscellaneous postsecondary schools. Total enrollment at these institutions was 3.36 million, which constituted about 7 percent of the entire Korean population. Enrollments in all types of higher education, excluding graduate schools, exceeded 62 percent of the total Korean population aged 18–21 years in 1996, while about

44 percent of this age cohort was enrolled at four-year institutions of higher education. Korean higher education has, therefore, already passed the threshold of being an elite system, and it has now become a mass system, in terms of the percentage of enrollments by the eligible age cohort.

Several factors contributed to the expansion of higher education, particularly among private four-year institutions in Korea. Examples include the continued rise in the birth rate in the decades since the Korean War and the social and national policy demands for a more equal distribution of educational opportunities by region, gender, and discipline. However, the most salient factor in the expansion of higher education had to be the traditional values of Koreans, who view four-year higher education as a valuable means for socioeconomic upward mobility. That was the reason that in Korea academe has been regarded as a growth industry and why private institutions have proliferated around the country. As table 7.1 reveals, private four-year colleges and universities significantly outnumber public ones, enrolling more than three-quarters of all students. In Korea, many entrepreneurs and leaders in private organizations and foundations have invested in establishing four-year colleges and universities. Some of these institutions have made considerable contributions to the development of high-level human resources in Korea. It was also true, however, that many private institutions were established primarily for the purpose of making profits from student tuition.

The quantitative growth in institutions and enrollments has been prodigious. There has also been a corresponding increase in the number of faculties and departments so that virtually all four-year colleges and universities have a full range of professional faculties. These expansions have affected the work environment of faculty at Korean colleges and universities.

A Changing Environment

The government of Korea has long assumed responsibility for the provision of higher education, directing attention and resources not only to public but also private institutions. The Korean higher education system has moved beyond the principle that the public national universities were supposed to serve the needs of the state while private institutions catered to the aspirations of individuals. This dualism assumed that the government was responsible only for national institutions and not private ones, particularly with regard to institutional finances. In reality, government subsidies for private institutions, which educate more than 75 percent of the country's students, have always been nominal but are now starting to increase. From the beginning, however, all institutions, private as well as public, have been under strict government control. Traditional government involvement in Korean higher education institutions includes enacting laws that require institutions to obtain government approval for all administrative policies and decisions.

For example, in Korea, a strict enrollment quota policy was instituted by the government, not only for public but also for private institutions. Every institution in the country attempts to increase its quota as much as possible during the government's annual assessment of enrollment quotas. In particular, a great many private institutions, whose financial stability is necessarily tied to enrollments, have expanded their business by increasing the number of buildings, departments, faculty, and enrollments, producing more tuition income.

It is widely recognized and predicted, however, that by 2003 higher education in Korea will be in a difficult period because of a downturn in the number of high school graduates. It is certain that the overall enrollment quota for higher education will then be larger than the number of eligible students. To make matters worse, the traditional preference for four-year institutions over two- or three-year junior vocational colleges is being challenged by the increasing focus on careerism in higher education. Many high school graduates want to get practical vocational training in two- or three-year junior colleges to join the work world as early as possible, avoiding four-year institutions, which are seen as less employment oriented. Many private four-year colleges and universities are suffering from low enrollments. The data for the Institution Survey conducted by Ministry of Education in 2000 (Ministry of Education, 2001) show that 33 percent of private four-year institutions are now enrolling less than 80 percent of their enrollment quota allowed by the government, resulting in depressed financial conditions and low expectations regarding their work environment on the part of faculty.

The traditional strong gravitation toward the capital city, Seoul, is increasing and has the potential to harm the academic environment. About 29 percent of all four-year colleges and universities are located in the Seoul metropolitan area, accommodating the same percentage of total national enrollments. Along with the job opportunities concentrated in the Seoul metropolitan area, admiration of city life and the presence of top-ranking institutions in Seoul have also influenced the preference of students for institutions in the Seoul metropolitan area. Many students enrolled at their local institutions seek to transfer to institutions in the Seoul metropolitan area. The concentration of the student population in the Seoul metropolitan area, along with the declining number of high school graduates mentioned above, exacerbates the financial problems facing local private colleges and universities. Many local private institutions may go into bankruptcy, raising questions about what faculty at these institutions will do for employment and where they might continue to pursue their careers as professors.

With declining enrollments, all faculty and other staff—except for those at a few elite institutions—are now forced to recruit students into their institutions. Recruitment might be called the fourth function required of professors, joining their traditional functions of teaching, research, and service. To attract more students and to survive the current financial constraints, institutions in all regions as well as in the Seoul metropolitan area have begun to accelerate change and the reform of their educational programs and student services. The primary emphasis has been placed on differentiating between the curriculum, academic management, student services, and the scholarship system. Many ideas have been developed and put into practice by private colleges and universities to build up their identities and their specific strengths and to restructure their operations on the basis of a cost–benefit analysis.

The need for institutional differentiation arises not only from the declining traditional college-age population and increasing student preference to study in the Seoul metropolitan area, but also from a shift in students' value orientation from intellectual to careerist goals. It is now clear that higher education must serve the diverse needs of students, who look to a college education as a route to professional career development. Four-year colleges and universities have begun to respond to these changes with more vocationally oriented majors, new programs, greater use of resources from the professional world, more flexible scheduling, affiliations with institutions abroad, and greater opportunities for experiential learning. Moreover, these alternatives are being developed by the full spectrum of institutions, from established and elite to marginal and unsophisticated. In contrast to the uniformity in the twentieth century, we now have numerous institutional and curricular models available in Korean higher education.

There is another condition that has led to changes in Korean higher education: governmental financial support for private higher education. The government has programs that provide direct and indirect financial assistance to higher education institutions and their students. In the 2000 academic year, funding for higher education accounted for only 9.9 percent of the government's entire education budget. Government financial support for private four-year colleges and universities did rise to 3.79 percent of total education expenditures in 2000, up from 1.39 percent in 1990 (Kim, 2001). Government support per student at private institutions is only equal to between 5 and 6 percent of that spent per student at public institutions. Increased government support for higher education institutions, particularly in the private sector, will be necessary in the future. It is still undecided as to how the increased funding will be distributed and in what amounts—one option is to continue to allocate government funding equally to each private institution, in proportion to its enrollment. This distribution plan, however, may not ensure the best use of the relatively small amounts received by individual institutions. For this reason, the government established a new funding policy for higher education institutions in the 1999 academic year.

The new policy, Brain Korea 21, was designed to elevate the quality level of Korean higher education in the twenty-first century to world class, in selected graduate schools and the natural sciences. The total amount of this subsidy, 1.4 trillion won (about U.S.$1.1 billion) for 1999–2005, is enormous in scale, especially when compared to previous government subsidies. This assistance targets selected public and private four-year colleges and universities and includes direct full financial aid to graduate students; specialized grants, such as for support of an applied physics research group; and grants targeted on the implementation of education reforms, such as restructuring department-based organizations into division-based ones. Thus, subsidies to private institutions vary enormously depending on the institution, field, and student. This policy supports some elite public and private institutions better than others, since enrollments tend to be more competitive than before. Not surprisingly, both students and faculty have been affected by such subsidy programs. Faculty have been affected dramatically, since they have to conform with the governmental reform guidelines—often against their will. These changes may pose additional threats to academic freedom and institutional autonomy and may also have an impact on academic ethics.

Student activism in Korean colleges and universities has been strong, historically. Since 1948, students have been involved in national economic and political affairs, but the event that initiated a period of unprecedented student activism was the 1960 April Revolution, against the dictatorial Rhee government. Since then, national attention has focused on students' flaunting of university policies and the violent tactics employed by government in response, which effectively closed down the universities. The period of student political activism led to the creation of the most effective of all student activist organizations, the General Association of Korean College and University Students. The democratization of Korean politics begun during the 1990s, however, has resulted in a decline in student participation and interest in political activism. Instead, student activism is focused on the academic institutions themselves. Students have gained increasing control over the academic core issues at most institutions, especially concerning tuition, curriculum, space and resource allocation, and other internal governance matters.

Student governments at colleges and universities have an official status. They are recognized and approved by the governing boards, specific powers are often delegated to them, and they always have an official budget. For student governments, the funds usually come in part from compulsory student fees collected by the institution. On the other hand, faculty have clearly been relegated to an increasingly marginal role in academic affairs. Faculty governance, however, is not yet well established at Korean colleges and universities. In the late 1980s, some leading faculty members began to urge the organization and recognition of faculty councils and faculty senates at each institution and direct elections for university president. The faculty council or senate has now become a vital channel of communication among academics. Despite government efforts to circumvent union activity, radical faculty proponents of faculty governance are developing new ways to protect the rights of their colleagues, including nationwide unionization of faculty and the creation of a strong association representing all college and university professors. Currently, the majority of faculty members seem indifferent to the unionization movement and hesitant to take an active part in it. When faculty unionization does come to campuses, however, it is likely that the faculty members and the administrator-trustees will take up adversarial positions. As members of bargaining units, faculty members may lose their individuality, flexibility, seniority, and personal relationships with individual faculty members and administrators.

The last salient feature of the changing environment concerns faculty evaluation. Many Korean faculty members at four-year colleges and universities are disturbed by the faculty "evaluations" that have been instituted in recent years. Some faculty members confess to a sense of apprehension, while others pay little attention to the process of evaluation. For a long time in Korean higher education, faculty were completely free from the burden of being evaluated by others. Once faculty were appointed, they could expect to be promoted and tenured over the course of a number of years. This independence was the reason why the academic profession has always been ranked as one of the best in terms of career track.

In the 1990s, the situation began to change. Many institutions of higher education began to advocate faculty evaluation and implement effective evaluation procedures. Now, all faculty members agree on the necessity for faculty performance evaluations, but no one has as yet developed an agreed-upon practice that works effectively. The intended purpose of faculty evaluations influences the kinds of information gathered, the sources of data, the depth of data analysis, and the dissemination of the findings. Faculty evaluations are supposed to improve faculty performance in research, teaching, and service and to enable the administration to make informed decisions concerning promotions and tenure, academic rewards—including sabbatical leave—provision of research funds, and rank. Each institution also develops its own system for evaluating faculty performance. Differences in how these measures are perceived and interpreted by administrators, faculty members, and even students often create conflict at individual institutions.

THE DEMOGRAPHIC PORTRAIT

A 2000 Ministry of Education survey collected data on a wide range of faculty and instructional staff, including permanent full-time faculty and temporary full-time or part-time instructional staff (Ministry of Education, 2001). "Permanent" and "temporary" refer to the appointment status of an individual. Permanent full-time faculty are engaged directly in teaching, research, related public service, institutional service, or a combination of these. They are tenured or can expect to hold tenure until the legal retirement age of 65. Temporary full-time or part-time instructional staff are employed for a fixed period and are prohibited from holding tenure. Such staff engage primarily in teaching and are paid by the number of hours they teach. They have no other responsibilities, nor do they have the right to take positions in the administration, serve as student advisers, or attend faculty meetings. They are not assigned any rank and are just called "professor" or "lecturer." These temporary instructional staff fall into three groups: adjunct professors, who are primarily employed at other nonacademic institutions; part-time lecturers, most of whom have no permanent jobs; and emeritus professors, who have retired from permanent university faculty positions.

Table 7.2 shows the number and distribution of permanent full-time faculty and temporary instructional staff at four-year colleges and universities in Korea in 2000.

Table 7.2 Faculty and instructional staff at four-year colleges and universities, 2000

Institutional type	Full-time faculty	Adjunct professors	Part-time lecturers	Emeritus professors
Public	11,359	289	10,281	1,139
Private	30,584	3,637	36,097	1,320
All types	41,943	3,926	46,378	2,459

Note: Data from Korean Education Development Institute, 2000.

Family Background

No data are currently available on the family background of faculty members in Korea. In the research undertaken for this project, the author surveyed a random sample of 92 faculty members from across the nation. The majority of these faculty members were from middle-class families; a good number of them were from upper-class backgrounds. It would be an exaggeration, however, to conclude that the professoriate is an aristocratic group. Few differences in family background were found among faculty of different disciplines except in the medical sciences, where a disproportionately high percentage of parents had degrees in the medical sciences. On average, 87 percent of all faculty had college-educated fathers, and 75 percent had college-educated mothers. It was also found that about 70 percent of the faculty came from families in which the fathers were in professional, managerial, or entrepreneurial occupations. Moreover, 25 percent of the sample had fathers who were in elite or high-status occupations. Also significant is that 14 percent of the faculty had fathers who had earned graduate degrees from colleges and universities abroad.

With regard to religious beliefs, the survey found that more than half of the faculty families were deeply or moderately religious, particularly in Protestant denominations. There were no differences in family religious commitment between faculty members at secular and nonsecular institutions.

Salary Level

In 1990, the average salary of full-time faculty members was $28,007 and in 2000, $31,727, as shown in table 7.3. Overall, average salary levels in 2000 were: full professor, $40,422; associate professor, $33,231; assistant professor, $28,948; and instructor, $24,305.

Table 7.3   Average faculty salaries by category and academic rank (U.S. dollars)

Rank	1990			2000		
	Public	Private	Total	Public	Private	Total
Full professor	31,150	36,638	35,043	39,037	42,628	40,422
Associate professor	25,427	31,049	29,455	32,298	36,037	33,231
Assistant professor	21,936	26,702	25,359	28,634	29,889	28,948
Instructor	19,552	23,194	22,171	24,286	24,332	24,305
All ranks	24,516	29,396	28,007	31,064	33,222	31,727

Note: Data from Korean Council for University Education, 2000.

Differences exist in average salary levels by type of affiliation, ranging from a high of $42,628 for full professors at private institutions, to a low of $24,286 for instructors at public institutions. On average, salary levels at private institutions exceed those at public institutions. From 1990 to 2000, the average salaries of faculty members increased by 26.7 percent at public institutions and by 13.0 percent at private institutions. Rapid salary increases continued until 2000, when the rate of increase began to slacken.

Compared to other professional and managerial occupations, academic salary levels are hardly impressive. In fact, the earnings of the highest-ranking faculty members are barely higher than those of workers in occupations such as engineering and business administration. By contrast, the salaries of academics at the lower ranks of assistant professor and instructor lag behind those of workers in comparable professional fields even though the latter typically hold only bachelor's and master's degrees.

In Korea, many faculty members supplement their faculty salaries with outside work—some of it performed during vacations, but much of it during their regular work week. Indeed, many institutions openly urge faculty members to engage in remunerative outside activities such as publishing; entrepreneurial activity; inventions; art; professional work, such as lecturing, consulting, part-time teaching at other institutions, and research; and miscellaneous moonlighting. Several studies indicate that academics in Korea express favorable attitudes toward their careers in terms of their earnings. A majority of faculty members indicate that they are satisfied with their salary levels and pleased with their career choice. A majority are ranked in the upper income levels of Korean society.

Gender

Up until the 1960s, both the faculty and the student bodies of four-year colleges and universities in Korea were overwhelmingly male. Over the past three decades, however, female enrollments at four-year colleges and universities have increased substantially. In 2000, women accounted for about 36 percent of enrollments at four-year colleges and universities and 35 percent of enrollments in graduate schools. In the case of the faculty, however, no such proportionate increases in female participation are discernible. In 2000, just over 13 percent of Korean faculty members at four-year colleges and universities were female. (See table 7.4.) Considering that 12.1 percent were female in 1990, the proportion of female faculty has held relatively constant with just a slight increase over the decade (Lee, 1992).

Female faculty representation varies across disciplines, from just 1.45 percent in engineering to 31.41 percent in the fine arts, music, and sport sciences fields. This distribution may be a function of the intersection of discipline characteristics with the dispositions of women, or with the academic climate of the disciplines. This pattern of concentration and scarcity of female faculty across the disciplines does clearly reflect patterns of gender and field distribution among doctoral degree recipients. Historically, women have tended to be disproportionately located at junior colleges (about one-quarter of all female faculty in 2000) and at the less-selective four-year colleges, and underrepresented in the university sector. The trend—particularly among "new hires" in recent years—appears to be toward a more equitable distribution across all institutional types. According to an analysis of Ministry of Education data, as presented in table 7.5, in 2000 women constituted 44.2 percent of faculty under age 30, while men comprised 55.8 percent; 23.3 percent of all instructors were

Table 7.4 Distribution of female faculty by field, 2000 (percentages)

Field	Female	Male	Total
Humanities	19.43	80.57	100
Social sciences	7.04	92.96	100
Natural sciences	15.99	84.01	100
Engineering	1.45	98.55	100
Medical sciences	15.66	84.34	100
Fine arts, music, and sport sciences	31.41	68.59	100
Teacher education	21.03	78.97	100
All fields	13.73	86.27	100

Note: Data from Korean Education Development Institute, 2000.

Table 7.5 Proportion of female faculty by age and rank, 2000 (percentages)

Age	Female	Male
Under 30	44.17	55.83
30–39	15.70	84.30
40–49	13.35	86.65
50–59	13.34	86.66
Over 59	9.18	90.82
Rank		
Full professors	11.98	88.02
Associate professors	12.66	87.34
Assistant professors	14.24	85.76
Instructors	23.25	76.75

Note: Data from Ministry of Education.

female, while 76.8 percent were male. This is evidence of a shift among new hires in recent years to create a more proportionate representation by gender. There is no evidence, however, that women faculty at Korean institutions are paid less than men, nor that they are promoted at a slower rate than their male colleagues. It does, however, appear that female scholars definitely have a lesser role in administration and governance, even after controlling for rank.

Age

The distribution of the faculty by age, at any given time, is a result of turnover—the entry of new faculty members and the departure of old ones. The process of faculty turnover at Korean four-year colleges and universities is in most cases simple and standardized. In theory, it is possible that entry and exit of faculty can occur at all ages from the 20s to the 60s, though there is a fixed retirement age of 65.

Based on an analysis of the age distribution of faculty in 2000, there seems to be a "graying," or aging, of the faculty in the fine arts, music, sport sciences, humanities, social sciences, and natural sciences. Faculty members are now clustered in the 40–59-year age group. In 2000, the mean age among full-time faculty members at four-year colleges and universities was 44.3 years, a slight increase from the 1983 average of 43.8 years. Over nearly two decades, the age distribution of the faculty, at least until quite recently, has been remarkably stable. Faculty have continued to be widely spread out over the various age brackets. Table 7.6 shows a moderate, but perceptible, aging of the faculty.

Table 7.6 The professoriate by age groups, 2000

Age group	Percentage
Under 30	0.3
30–39	24.6
40–49	45.6
50–59	21.8
Over 59	7.7
All ages	100.0

Note: Data from Ministry of Education.

While the aging of faculty across Korea has raised concerns over the potential effect on teaching quality, no serious problems have as yet exhibited themselves. However, the decline in the number of student enrollments means that fewer young people will be recruited into the academic profession and that the profession will become more heavily populated with older faculty.

It is widely believed that having faculty representing a wide range of age groups is a good thing, particularly for the purpose of maintaining and increasing the internal cohesiveness of departments. Many faculty members believe that conflicts are apt to occur more frequently among faculty members in the same age cohort, which is the very reason why some departments have instituted a policy of seeking to recruit new, young faculty.

Rank

In recent years, one issue of much concern regarding undergraduate education is the widely held belief that there are too many senior faculty members at four-year colleges and universities. Since senior faculty are not required to undergo evaluation, the fear is that the lack of oversight will result in a decline in teaching vitality and productivity, due to a lack of challenges. Table 7.7 shows the top-heavy pattern of faculty distribution by rank. In past decades, the Korean professoriate has been divided fairly equally among the four ranks. However, since the 1990s, as the tenure system has been widely adopted, the percentage in the two upper ranks has risen and the percentage in the two lower ranks has declined. In 1983, the ratio of the upper two ranks to the lower two ranks was 0.86, but that changed to 2.16 in 2000.

While one cannot determine for certain if this top-heavy structure is a positive or negative thing, there is evidence of detrimental effects

Table 7.7   The professoriate by rank, 2000 (percentages)

Rank	1983	2000
Full professors	27.6	44.4
Associate professors	18.5	23.9
Assistant professors	29.7	21.9
Instructors	24.2	9.8
All ranks	100.0	100.0

Note: 1983 data from Lee, 1983 and from Ministry of Education.

on higher education. One impact pertains to money. Because salaries are based on seniority, the relative increase in the proportion of senior faculty has raised the operating costs of colleges and universities. Therefore, at some institutions, a new policy has been introduced that puts faculty salaries on an individual basis, rather than on the basis of rank.

Another issue is the fact that high-ranking faculty (full and associate professors) are often not subject to the rules of tenure and faculty evaluation. It is true that in recent years many institutions have begun giving tenure to faculty members once they are promoted to the rank of associate professor, but a few institutions only grant tenure at the rank of full professor. It has now become more universal that tenured faculty are no longer evaluated and receive automatic salary increases each year. Critics of this system predict that such a policy will undermine the vitality and productivity of faculty, as well as the finances of institutions. Thus, it is thought that all faculty, irrespective of rank, should be evaluated annually and their salaries raised on the merit of their achievements.

Academic Qualifications

Over the past several decades, the education levels of Korean faculty have changed significantly. Until the 1970s, their formal education seldom advanced much beyond the master's level. Since the mid-1980s, however, with the gradual popularization and professionalization of graduate education worldwide, faculty members are increasingly being directed into many specialized fields. The doctoral degree has become the norm for faculty at four-year colleges and universities.

The percentage of all faculty with doctorates has risen substantially, from 39.8 percent in 1983, to 73.5 percent in 1990, and 83.3 percent in 2000 (Lee, 1992). Correspondingly, the percentage with only a bachelor's degree has diminished to almost zero. Table 7.8 shows the

Table 7.8   Degrees held by the professoriate, 2000 (percentages)

Degree	1983	2000
Doctorate	39.8	83.3
Master's	52.9	14.8
Bachelor's	7.3	1.9
All degrees	100.0	100.0

Note: Data from Lee, 1983 and Ministry of Education.

Table 7.9   Doctoral degrees, by place, 2000 (percentages)

Place	1983	2000
Korea	61.0	58.9
Abroad	39.0	41.1
Total	100.0	100.0

Note: Data from Lee, 1983 and Ministry of Education.

percentage of faculty members with doctorates in 1983 and 2000. Faculty experienced a significant rise in professionalization with the increase in the number of doctoral degree holders by 2000. In the natural sciences and engineering, more than 95 percent of all faculty hold doctoral degrees, compared with 86 percent in the humanities and social sciences and 76 percent in the medical sciences. In the fine arts, music, and sport sciences, however, only 30 percent of faculty hold doctorates. In interpreting this data, one might surmise that the doctorate is not always necessary or appropriate for faculty in all disciplines.

The proportion of faculty who have earned their doctorates abroad has been steadily increasing in the natural sciences and engineering, but has decreased in the humanities, social sciences, fine arts, music, sport sciences, and medical sciences. In the aggregate, there was a slight increase in the number of faculty with doctorates from abroad, as shown in table 7.9. It is also notable (see table 7.10) that the proportion of faculty who have earned doctorates in the United States increased from 56.4 percent in 1983 to 67.6 percent in 2000. In particular, in the natural sciences and engineering, more than 70 percent of faculty with doctorates from abroad did their advanced graduate study in the United States.

This increase in the proportion of degrees earned in the United States can be attributed to a number of factors: the universally

Table 7.10 Faculty with degrees earned abroad, 2000 (percentages)

Country	1983	2000
United States	56.4	67.6
Other countries	43.6	32.4
Total	100.0	100.0

Note: Data from Lee, 1983 and Ministry of Education.

recognized quality of an American graduate education, especially in comparison to Korean graduate education; the fact that professors with U.S. doctoral degrees tend to encourage their best students to study in the United States; the additional credits awarded applicants holding American doctoral degrees when an institution recruits new faculty members; the increasing prosperity of globalization in all academic fields; and the broad government support for studying abroad.

Some may jump to the conclusion that the overrepresentation of American-produced scholars in Korean academe might diminish the credibility of Korean academics. It would be a mistake, however, to conclude that the Korean academic system has become overly dependent on American higher education.

ACADEMIC APPOINTMENTS

In recent years, many colleges and universities in Korea have studied faculty appointments, focusing in particular on the criteria used in the evaluation process. This section uses those institutional studies but is based primarily on the author's years of experience in assisting many colleges and universities to develop successful faculty evaluation programs, as well as in evaluating faculty members, as an administrator at the top-ranked private research university. This section's focus is practical rather than theoretical. It offers pragmatic information about the current state of academic appointments and due process at Korean colleges and universities, including recruitment and initial appointments, reappointments, promotions, tenure, and retirement.

The Recruitment Process

In Korea, no college or university has ever completely filled the number of legally required faculty positions, which is determined by government in proportion to student enrollments. In 2000, only 73 percent of the authorized positions were filled at public four-year colleges and universities and 65 percent at private ones. Government regulations currently require that institutions must have one faculty member per 25 students in the humanities and social sciences, one per 20 students in the natural sciences and in engineering, and one per 8 students in the medical sciences.

Consequently, there are many staff vacancies and the number continues to increase with the expansion of enrollments at all institutions. In addition, vacancies in faculty positions are also caused by retirement, noncontinuation of contracts, and illness or death. There is very little voluntary turnover, however, and mobility among institutions is rare. Increases in budgets for faculty recruitment have not kept pace with the number of vacancies that need to be filled, negatively affecting institutional capabilities to hire faculty.

The central administration at most institutions has the power to set the number of positions to be filled. Each department is asked to determine the number of positions (and in which areas) that need to be filled and to submit their findings to the central administration, which then allocates faculty to departments. Most institutions seek to fill vacancies on a regular basis, often twice a year. Academic positions are usually advertised in daily newspapers and a weekly chronicle for higher education, as well as on the institutions' Internet homepages. Announcements are often sent to other institutions as well.

The Screening Process

The central administration collects, reviews, and evaluates the applications. Evaluation procedures vary among institutions, but the usual practice is a three-stage screening process.

The first stage of screening occurs at the departmental level. Typically, a screening committee is made up of five to seven faculty members, who are appointed by the department chairperson. At many institutions, all faculty of a department are encouraged to participate in the evaluation of candidates, but the decisions are usually made by the committee. The functions of the screening committee include reviewing and evaluating applications, including academic records and published research; interviewing the candidates and in some cases arranging their oral presentations or lectures; and making final decisions and recommending two or three persons for each position, in rank order, to the dean or division head.

At the first stage of screening at the department level, a wide range of criteria are used for evaluating the competence of an applicant. There are no precise formal standards for admission to the professoriate. Each institution is free to choose its faculty according to its own judgment, without external or governmental regulation. As the basic standard, the Ph.D. is the most frequent credential for faculty at four-year institutions. Holding a doctoral degree, or its equivalent, however, is on its own not a sufficient qualification for membership in the professoriate. Many institutions accord greater importance to academic research experience and publications, the reputation of the institution from which the applicant earned his or her degree, academic achievement records, teaching ability, and personal attributes. Sometimes unexpected requirements for employment may arise. For example, some institutions focus on English proficiency as an indication of faculty competence. It is also not uncommon for local institutions to require that selected applicants move their place of residence to the vicinity of the campus.

The second stage of screening consists of the dean's recommendation. Once the departmental committee determines their recommendation, they send their decision, with their evaluation files, to the dean of the college or division. The dean organizes another evaluation committee made up of department chairpersons and associate deans. The committee reads the evaluation files to decide how well and fairly the department committee has followed the evaluation procedures. If the dean's committee discovers inappropriate or unfair evaluations or recommendations, it will return all files to the department committee and ask that the candidate be reevaluated. If the files seem to be in order, the dean's committee may then recommend the department's selection to the central administration committee, which then makes the final decision on the appointment of an applicant.

The third stage of screening involves the final decision on the part of the central administration. At Korea's public institutions, the legal power to appoint faculty members resides with the president or the ministers of education and at private institutions with the director of the board of trustees. In reality, the power is actually legally delegated to the president of each public or private institution. This means that college and university presidents have strong powers to appoint, reappoint, grant tenure to, and even fire faculty members. At most, but not all, institutions, however, the president works in a democratic and collegial fashion, by delegating the hiring authority to a committee that makes the final decisions on new faculty appointments. This committee, which is appointed by the president, consists of five to ten high-ranking administrators and two or three faculty representatives and is chaired by a vice president or a dean of academic affairs. The primary responsibilities of the central administration committee include reviewing all evaluation procedures made at departmental and collegewide levels to ensure fairness, examining the recommended applicants in rank order and making the final choice, and deciding the rank and type of appointment for the selected applicant. If any irregularities are found in the evaluation and recommendation procedures in the lower stages of the screening process, the committee may return the files and decline to make a final decision. In addition, the central administration committee occasionally decides that a position for which candidates are being screened does not meet immediate institutional needs. On these occasions, the central committee reallocates the faculty opening to a department that is deemed to have greater need.

Some Observations

Under current conditions, the academic profession is highly attractive to persons who are talented and have earned advanced degrees. Most graduate students believe that even after many years of advanced study, getting a full-time job in academe is highly competitive and difficult. Nevertheless, they remain strongly inclined toward the academic profession and willing to make aggressive efforts to cultivate their talents as scholars and college teachers and thus improve their chances of earning a faculty position. Accordingly, when a full-time faculty position opens up, the competition is remarkably keen. Thousands of part-time faculty with Ph.D.s who have spent years waiting for a full-time position rush to apply. In the 2000 academic year, a total of 2,559 new full-time faculty members were hired at four-year institutions of higher education in Korea. It is estimated that more than 12,000 people applied for those positions. The highly competitive nature of faculty recruitment has created a number of problems in academe.

Perhaps the most elusive dimension of faculty appointments concerns the integrity of the selection process. There are a number of practices that raise questions about fairness in the faculty selection process. The first of these concerns the announcement of a position that is in reality to be filled with a preselected candidate. This is an extremely deceptive practice but one that happens quite often and involves behind-the-scenes prearrangements with certain faculty members. The second of these questionable practices is the fact that some faculty who have a great amount of power in the selection of new faculty take cash bribes from applicants. Third, faculty members often have a strong preference for applicants whose degrees were earned from the same institution as theirs. In fact, more than 70 percent of the faculty at leading research universities in Korea have bachelor's degrees from the institution where they work. Fourth, the increasing abuse of power by the president or chairperson of the board of trustees—in the form of placing their own people on campus—has become particularly problematic at some private colleges and universities.

Reappointments

In Korea, it is now a common practice at all four-year colleges and universities that new faculty are appointed for a fixed period of one to three years, regardless of their initial rank. Newly appointed faculty are put on a kind of probation or "reappointment" track, which when completed allows them to move up the career ladder, in terms of rank and salary, and to earn tenure. This reappointment process can also lead to a form of contract employment for a fixed period, unless and until the faculty member is tenured. The contract period varies by rank as well as by institution—three to five years for assistant professor and four to seven years for associate professor. If faculty are not reappointed within the fixed contract period, they must leave the institution.

Each institution develops strict evaluation procedures and standards for reappointment that mirror the three-stage process for initial appointments. However, the evaluation of faculty performance and the criteria for reappointment are quite strict and comprehensive, covering the quality of classroom teaching, research, and service.

Two significant observations can be made with regard to the practice of reappointments. The first concerns the number of faculty members who are *not* reappointed. In 2000, a total of just 29 faculty members at four-year colleges and universities nationwide were denied reappointment. Thus, almost 100 percent of faculty are being reappointed, which calls into question the relevance and validity of the whole process. The other relevant observation is that faculty members who take issue with their failure to be reappointed attribute it not to low productivity or other academic inadequacies but to the fact that they were perceived as provocative by central administrators, department colleagues, or the board of trustees. The reappointments process has in fact been used by some private institutions to rout out and remove troublesome faculty members, making it often hard to be certain why some faculty are not reappointed. During the period of military rule, the government also began to abuse the process, ousting faculty members who were leaders in the antigovernment movement.

Promotions

The most immediate concern of faculty, beyond salary, is achieving promotion in rank, which is of consequence to academics in Korea, just as it is for military personnel, for example. The Korean higher education system contains four academic ranks: instructor, assistant professor, associate professor, and professor. As mentioned earlier, at most four-year institutions, holding a doctoral degree is a prerequisite for initial appointment at this rank—except in the arts, music, and a few other specialties.

At most Korean four-year institutions today, an academic career begins at the assistant professor rank. Assistant professors are usually given a five-to-six-year trial or probationary period, during which they need to establish a record of achievement to warrant being retained indefinitely. At some institutions the individual's progress toward this goal is reviewed each year; at others, review does not take place until the end of the fifth or sixth year. If the decision is favorable, the assistant professor is promoted to associate professor, usually with tenure. In some cases an individual may be promoted to associate professor without tenure. If the decision is unfavorable, the assistant professor is offered an additional year before undergoing another review; if the decision is again negative, the individual is terminated. Termination is decided on reluctantly, however, and it is usually postponed to the very last moment. Associate professors, even those with tenure, undergo another strict examination before being promoted to the rank of full professor. At most institutions, associate professors can ask to be reviewed after four to six years. In most cases, the process for assistant professors is identical to that for associate professors.

In 2000, the average promotion rate from assistant professor to associate professor was 63 percent and from associate professor to full professor, 73 percent. At some elite research institutions, however, the rates of promotion were far lower than those at other institutions. For example, the rates of promotion at some universities are as low as 22 percent from assistant to associate professor and 34 percent from associate to full professor. There were no significant differences in promotion rates at public and private institutions. At the more prestigious institutions, however, it often takes longer to be promoted to the higher ranks. Faculty at prestigious institutions are also less likely to be promoted or are likely to be promoted more slowly. On average, it takes 5.3 years to be promoted from assistant to associate professor, while promotion from associate to full professors takes an average of 4.9 years.

As noted earlier, in Korean higher education promotions are based on three primary criteria: teaching, research, and service. Competence in teaching has traditionally been the most elusive attribute to measure. In Korea, the standard teaching load for professors is nine credit hours per week at most institutions and six hours at a few others. Assessing faculty teaching competence takes into account course load, student teaching evaluations, course development and instructional innovation, advising of graduate students at the master's and doctoral levels, as well as other activities. It should be noted that there is a continuing debate in the academic community as to how much weight should be given to student evaluations, since they are influenced by factors such as grading policies or the personal charisma of professors. Notwithstanding the imprecision of such measures, most colleges and universities have begun to make formal attempts to obtain some index of informed consumer response to their faculty members. This is customarily done through carefully constructed questionnaires that the students are asked to complete toward the end of a given course.

Compared to teaching performance, a professor's productivity in research and scholarship—the second criterion—is relatively easy to measure. At the most basic level, one sees the simplistic quantitative approach that so often typifies the less-sophisticated faculty evaluation committees. It is, however, generally understood that the number or length of published books and articles may not be the only accurate measure of research performance. At most institutions, additional questions are asked concerning faculty publications: in what type of journals has the person's work appeared (refereed journals, Korean journals, international journals, SCI-registered journals), and were the articles in question single authored or coauthored? If the latter, was the individual the primary author? If an article was originally a paper read at a conference, was the conference program refereed? Was the paper an invited rather than unsolicited submission? How often has the article been cited by other scholars? Regarding books, similar questions are asked. Was the book a textbook, professional research-based book, monograph, translation, edited book, collection of essays for general publics, or a contract research report? Who was the publisher, and what is the scholarly reputation of the publisher? Was the book single authored or coauthored? How many grants has the person received, in what amounts, and from what sources? Has the individual obtained any patents, awards, or prizes? In these ways the evaluation committees are able to utilize scholarly peer reviews, sometimes even going to the trouble of soliciting ad hoc letters of review from scholars or experts in the field.

A third main way of evaluating faculty is to examine their levels of internal and external service. Internal service includes participation at various levels of administration, work on committees and councils, participation in the screening and selection of new students, and other institutional housekeeping activities. External service includes work in connection with the community, government, and national and international organizations and institutions; participation in academic societies and associations; and some activities involving mass media. This type of faculty activity presents complicated and unfamiliar problems of assessment and evaluation. With the increasing awareness of the importance of public relations for higher education, however, institutions are developing effective measures to document faculty service for inclusion in promotion decisions.

Taking the three criteria collectively, it is to be expected that an individual faculty member will have varying degrees of competence in these areas. This raises one important question: Should the three criteria be assigned equal weight in the final evaluation? Institutions differ in their answer to this question, with many relating the relevance of each area to the mission of the institution.

Evaluations and decisions about promotion and tenure are customarily made through the three-stage process mentioned earlier. In contrast to initial appointments, however, the final stage, involving a high-level committee chaired by a vice-president or the president, is the most decisive one in the promotions process. The department chairperson and the dean usually support the promotion of their faculty members, but the central administration committee reviews their recommendations most thoughtfully.

Tenure

The Korean system of higher education has instituted an academic tenure system that is based on the American system. In the 1990s, as part of the higher education reform movement, many institutions of higher education in Korea began to implement a tenure policy as a means of ensuring academic freedom and improving faculty productivity—goals seen as essential to fulfilling the purposes of higher education.

At first, many institutions offered tenure, without additional evaluation, only to full professors, while some institutions offered it to both full professors and associate professors. Today, most institutions have formal tenure policies and procedures that provide explicit

Table 7.11   Faculty and tenure, by academic discipline, 2000 (percentages)

Field	Tenure	Tenure track	Total
Humanities and social sciences	62.1	37.9	100.0
Natural sciences	68.5	31.5	100.0
Engineering	56.1	43.9	100.0
Fine arts, music, and sport sciences	62.4	37.6	100.0
Medical sciences	51.0	49.0	100.0
All combined	61.2	38.8	100.0

Note: Data from Korean Council for University Education.

guidelines concerning qualifications and criteria for reappointment and the awarding of tenure.

As shown in table 7.11, academic fields vary considerably with regard to tenure at four-year colleges and universities. In total, over 60 percent of full-time faculty have tenure, which is less than the percentage of full and associate professors presented in table 7.7. The remaining 39 percent of faculty members have tenure-track appointments. These are full-time faculty members who have probationary status and are under consideration for tenure. Some are young faculty seeking their first tenured positions; the others are older faculty who have failed to gain a tenured appointment and are trying again.

Even before the tenure policy was widely accepted by institutions and legally required by the government in the 1980s, it was extremely difficult to terminate any individual even after the fixed contract period of appointment. Today, however, all institutions seem to have a firm tenure policy, through which faculty members promoted to full professor or associate professor in some institutions are automatically given tenure. In spite of strict evaluation procedures for promotions, no one is known to have been dismissed due to an unfavorable decision on promotion.

The tenure system is necessary for the protection of the academic freedom of individual faculty members. Equally important, tenure is essential to protect the intellectual autonomy of universities as well as individual faculty members. At Korean institutions, tenure is instead viewed in economic terms—principally as a device to protect job security. In some cases, it has unfortunately been turned into a protection for the weak—the plodding, unresourceful faculty who would find it difficult to get rehired if placed on the open academic market. Because of this unusual situation, some opposition to tenure has begun to develop. Institutions are beginning to evaluate the ratio of tenured to nontenured professors. A number of tenured professors, in the absence of any regular evaluations, lack the incentives to remain relevant or to improve their teaching, research, and service. Each year, their salaries increase, without any restrictions or review. That is the reason why an elite private institution like Yonsei University has adopted procedures for evaluating all the tenured professors, to determine promotions in their pay scale. An important task for improving future tenure policy in Korean higher education is to confirm the principle that tenure should not be awarded automatically along with promotion to associate or full professor position, but only in recognition of superior performance.

Retirement

It has been standard practice at all four-year institutions of higher education to fix the mandatory retirement age at 65. Among the provisions for retirement, all institutions have established retirement annuities accumulated throughout the working life of each individual faculty member. These annuities are payable after retirement in installments throughout the remaining life of the retiree and of his or her spouse. The annuity plans are compulsory for all full-time faculty members at both public and private institutions and require regular monthly contributions by both individual faculty members and their employing institutions.

At many colleges and universities additional part-time service has been permitted, under the title of "emeritus professor." Not every retired professor is given the title, however. To be selected as an emeritus professor, a professor must have been employed, by the same institution, for more than 20 years at some institutions and for 25 years at other institutions. In a few cases, the professor must obtain the approval of faculty members in the department, and he or she is also officially evaluated and selected through the same three-stage evaluation process as in initial appointments and promotions. The most beneficial advantage of being an emeritus professor is the title, which allows the individual to remain professionally active and have access to campus facilities.

In addition, many institutions adopt early retirement plans for faculty members who have served for more than 20 years but who want to leave their institutions for a variety of reasons. The number of early retirees is now increasing, and the trend is well accepted—particularly because the more early retirees there are, the more openings there will be for new Ph.D. holders to be recruited to the institution. Of course, early retirees are given retirement annuities with additional monetary compensation.

Faculty members often make more once they have retired at the mandatory age. Evidence suggests that most faculty continue their professional activities, engaging in some form of teaching, research, or service during the period after retirement. A few inactive retirees, who may have done less planning for their retirement, abandon their professional contacts.

ISSUES AND CHALLENGES

The Academic Labor Market

The academic labor market consists of institutional demand for faculty staff members and the supply of educated persons, particularly doctoral degree holders seeking professorial jobs. The number of faculty positions to be filled is determined primarily by student enrollments and by an institution's workload and responsibilities. It is also affected by the number of replacements needed for persons departing from academe through retirement, death, or resignation. On the other hand, the supply of faculty is drawn from people who are attending graduate schools in Korea and abroad and who seek to enter the academic profession. Other candidates for faculty positions work outside academe in a wide range of scholarly, scientific, and professional fields, but have strong inclinations to enter academe.

In the current depressed period in Korean higher education—and perhaps on into the future—the corps of young persons preparing for the academic profession far exceeds the demand for academics. More than 13,000 talented young scholars who currently hold Ph.D. degrees are not employed in academe. Many work as poorly paid, part-time lecturers. According to a report from the Ministry of Education, between 2002 and 2006 more than 40,000 doctorate holders will seek to join the academic labor force, but many of them will be forced to accept part-time college teaching or research positions with temporary appointments (Ministry of Education, 1999). In the past, colleges and universities in Korea employed local professional people such as practicing lawyers, doctors, researchers, and accountants as part-time instructors. They also employed new doctoral degree holders in full-time appointments. All this has now changed, however. The heavy use of part-time instructors in recent years is directly related to the ongoing financial difficulties facing institutions. Accordingly, when the financial stability of higher education becomes increasingly precarious, part-time instructors may be hired in larger numbers to reduce instructional costs, continuing to weaken the job market for potential full-time faculty. This growing dependence on part-time faculty is now widely regarded as a serious problem for academe, particularly in terms of the quality of instruction and waste of highly competent talent.

One may ask why doctoral production continued at a high rate despite the worsening academic market. One reason was the national economic depression in the 1990s. Many talented college graduates could not find jobs, so they simply continued their studies. Another reason is that government policy has encouraged young people to study abroad, opening a door that had long been closed. Still another factor is that the women's movement encouraged and emboldened many young women to seek Ph.D.s, resulting in women entering the academic labor market in unprecedented numbers.

In the future there will likely continue to be this serious one-sided growth in the pool of highly educated people without a corresponding number of faculty positions. The result will be stiff competition between new Ph.D.s and established members of the professoriate for the limited number of faculty positions throughout the country.

Faculty Evaluation

At Korean institutions of higher education today, the main purpose of faculty evaluations is to improve performance and to provide the rationale for administrative decisions on tenure, promotion in rank, and retention. Faculty members generally seem to agree that faculty evaluations are necessary, and in the future, performance evaluations will likely become common practice at many institutions. There are, however, strong pockets of resistance and resentment among professors about these faculty evaluation programs. Korean professors often feel overwhelmed by the multiple sorts of evaluations to which they are subjected: institutional evaluations, department evaluations, graduate education evaluations, government-funded research performance evaluations, teaching performance evaluations, central administration evaluations for budgeting, and so on. Some faculty members argue that all these evaluations contribute to an academic atmosphere of suspicion and anxiety, and may even impede the quality of faculty performance. It is often the case that evaluations of faculty performance at four-year colleges and universities are done simply for the convenience of administrators, not for the benefit of faculty members or to improve the quality of the education students receive.

Ideally, all faculty members should accept evaluation programs, show confidence in the programs' integrity, and take an active part in building due process into academia and safeguarding the legal rights of professors in tenure, promotion, and retention decisions. In the future, with open evaluation procedures, no professor could be removed by an authoritarian president or chairperson of the board of trustees, particularly at private colleges and universities. Accepting that faculty evaluation is a complex and evolving process, faculty members must work together to develop the criteria, standards, and evidence needed to maintain fairness, efficiency, and effectiveness. Nevertheless, faculty evaluation has never been and will never be a panacea to cure all maladies in academe but is a useful tool to address one area of concern.

Teaching Environment

At a department orientation for undergraduate freshmen, a professor made a brief speech: "I will never try to remember you by name. I cannot be kind to you. Don't expect any advice from me. I just welcome the intellectual challenge you provide me...." The role of the professor is changing from that of the traditional generalist to the academic specialist, in response to changes in the nature and quality of students, faculty peers, and administrators, and in the expectations of institutions as well as of society. In particular, forces other than diversity and growth in student numbers are producing changes for the academic workplace. New intellectual and social needs, stemming from the large and rapid expansion of knowledge and information technology, the discovery of new theories and methods for organizing and interpreting knowledge, and the ever-closer relationship between the campus and society have all made an impact at institutions of every kind and size. As a result, there have been changes in curriculum, in concepts of education and particularly liberal education in the undergraduate years, and in teaching approaches.

It is no longer sufficient for professors in Korea to be rigorously trained scholars; they must also be motivated teachers. This need will only become more acute, given the low priority afforded to teaching at many colleges and universities, especially at a time of sharply declining enrollments.

Institutional Diversity

Institutional diversity in higher education is critically important, because it not only meets institutional and societal needs more effectively but also because the differentiation of component parts leads to stability that can protect the system itself. It is widely believed that diversity is the essential and primary means through which Korean higher education institutions evolve. During the period of over-enrollments, the survival and stability of institutions of higher education were not genuine concerns. The situation now is quite different. Institutions in Korea are differentiated only by the average college entrance examination scores of the students they admit. It is quite difficult to find any other significant differences in curriculum, organization, governance, funding, or, even, in expressed goals. Over the next ten years, however, one can expect to see significant changes in the Korean higher education system as competition for increasingly scarce resources intensifies. Most likely this competition will lead to institutional instability and closures.

The maintenance and enhancement of institutional diversity are critical to the future stability and responsiveness of Korean higher education. It is imperative that each faculty member be an agent for change in creating, enhancing, and maintaining the diversity of his or her institution. There are several ways to encourage diversity and variability. For example, institutions can focus on curriculum development and institutional mission in response to student needs in the changing social and economic environment. Faculty should play a key role not only to ensure their own survival but also to ensure the survival of their institutions and the entire Korean higher education system.

Ethical Responsibility

In the West and in the East, colleges and universities have historically served as a forum for the development and sharing of ethical ideas and ideals. Today, institutions are experiencing an explosion of interest in integrating moral and ethical considerations into virtually all academic disciplines. It is ironic that colleges and universities have largely refrained from raising crucial ethical questions about their own principles, policies, and practices. While professors in Korea have been critical of others, they have deflected criticism of themselves by hiding behind claims of academic freedom. The time has come for serious inquiry into and scrutiny of the faculty's and institution's own ethical postures and behaviors.

Ethical dilemmas confronting professionals in Korean higher education today include a variety of personal and professional issues such as engaging in supplemental income-generating activities, sexual harassment, plagiarism, and the abuse and exploitation of student workers. In fact, truly unethical behavior is not uncommon among academicians in all aspects of their activities and campus life. High standards of behavior should be deemed fundamental on campuses. Grades, graduation, and faculty appointments should no longer be provided in return for cash or favors. In these perilous times, success of a college or university depends largely on institutional credibility, and nothing bankrupts credibility more promptly than the suspicion of ethical dereliction.

Conclusion

This chapter has examined the major developments involving the Korean academic workplace. Careful study of the issues yields a number of conclusions and recommendations.

To begin with, the academic profession is reasonably sound and healthy. On the whole, members of the academic profession are intelligent, well educated, hard working, productive, and honorable. They place a high value on autonomy, academic freedom, and collegiality. In general, they are satisfied with their salaries—a fact that helps to maintain morale in the academic profession.

The Korean academic profession may, however, no longer be stable. In many respects, the condition of the academic workplace seems to be slowly deteriorating. This trend has, of course, been more of a problem at some institutions than at others, but few have gone unscathed. The working conditions in Korean higher education are far from ideal and could worsen.

Many colleges and universities have become huge and impersonal institutions, with many administrative layers. Moreover, the quality of education being provided is not what it should be. The adjustment of academic programs and methods to the needs of their new clientele is still incomplete. The excessive concentration on job-related studies at the expense of liberal education compounds the problem of providing quality education. Unemployment rates among Ph.D.s have remained remarkably high. The market has been glutted with unemployed Ph.D.s, and the proportion of part-time faculty has increased dramatically. The increase in the number of full-time professors has now slowed, and evaluations of faculty performance have become more rigorous. Job security has become less certain than it was in more prosperous times, and this trend, particularly at some private institutions in Korea, has in turn impaired academic freedom in subtle ways. Looking ahead, the peril of these trends for Korean higher education appears ongoing.

The long-predicted decline in the youthful population still lies ahead. The political and financial conditions of the nation are still not stable enough to support larger government appropriations for higher education. These conditions will continue to lower the commitment of academic professionals to serve students and society. Many private institutions will turn to profit-making business enterprises to supplement their finances. The tragedy of such a trend is that it will undermine the institutions' primary missions—teaching, research, and public service—that have long characterized Korean higher education.

An overwhelming majority of institutions, particularly private ones, are suffering financial difficulties. To adjust to these increasing difficulties, many colleges and universities need to develop new fiscal approaches to support and maintain quality higher education. As is now occurring at some institutions, it may be desirable to pursue a strategy of selective excellence in order to maximize fiscal resources. In other words, many believe it is better to focus efforts on chosen areas, rather than to spread scarce resources too thinly in a futile attempt to be everything to everyone. The strategy of selective excellence may emerge as one of the most significant trends in Korean higher education in the coming decades.

Professorial life in Korea is still attractive to its practitioners. However, any profession afflicted by both financial challenges and a perceived deterioration of the work environment could soon have difficulty in attracting and retaining highly talented persons. Though not immediately evident, this is the problem with the most far-reaching consequences: Korean higher education may find itself with an inadequate instructional staff, and that will exact heavy social costs. To attract and retain an adequate supply of the brightest and most talented persons for the academic profession requires working conditions that are competitive with those of other professional occupations that hire comparable talent.

Several issues regarding faculty contracts are emerging in Korean higher education: appointments, reappointments, promotions, tenure, compensation, salary level, collective bargaining, and retirement. These elements of the academic profession have been undergoing serious scrutiny. The element that has received the most intensive criticism is tenure, and how the debate is resolved will be central to establishing a satisfactory work environment for the academic profession. The only real hope for maintaining a highly effective academic workplace is for faculty members, administrators, government officials, and concerned citizens to redouble their efforts to reorient their thinking about higher education.

References

Kim, Sukhyun. (2001). *Urinara Daehak Jaejung Siltaewa Kunsilhwa Jonryak* [Current Situation of Korean Higher Education Finance and Strategy For Enrichment]. *Daehakkyoyook* [University Education], 109, 59–67.

Korean Council for University Education. (2000). *University Development Index in 2000.* Seoul: Korean Council for University Education.

Korean Education Development Institute. (2000). *Statistical yearbook of education, 2000.* Seoul: Korean Education Development Institute.

Lee, Sungho H. (1983). *Jonkook daehakkyosoojawonui teauksung yunkoo* [An analysis of the characteristics of college and university professors]. Seoul: Korean Council for University Education.

———. (1992). *Hankookui Daehakkyosoo* [Academic professoriate in Korea]. Seoul: Hakjisa.

Ministry of Education. (1999). *Kyoyook baljeon 5 gainyun geihoik* [5-year plan for educational development]. Seoul: Ministry of Education.

———. (2000). *Survey of independent graduate schools.* Seoul: Ministry of Education.

———. (2001). *2000 institution survey.* Seoul: Ministry of Education.

8

The Fall of the Guru: The Decline of the Academic Profession in India

N. Jayaram

In the land of the guru, the teacher (*acharya*) was once upon a time venerated as a demigod (*deva*). From the village schoolmaster to the university don, in India teachers have traditionally been accorded the highest esteem, even if it was not matched by commensurate economic rewards. However, over the past few decades all this has changed: The academic profession has experienced a precipitous decline, and the once-revered guru has fallen. As if a reflection of this, the eulogistic description of teaching as a "noble profession" has become conspicuous by its absence in the current discourse on the subject.

The coming crisis of the teaching community was highlighted in the late 1970s by a compilation of essays (*The Indian Academic Profession*) covering the various facets of the academic profession in the country (Chitnis and Altbach, 1979). Significantly, this critical appraisal, the first serious attempt of the sort in India, was undertaken during the period of phenomenal expansion of higher education. Using this compilation as a benchmark, this chapter examines how the crisis in higher education has worsened since then and how the profession has fast become moribund.

THE CHANGING CONTEXT OF HIGHER EDUCATION

After a long period of protected expansion with state patronage until the mid-1980s, higher education in India has entered a period of stunted growth and uncertain future. A complex turn of events has thrown higher education into a vortex of change, inevitably affecting the teaching community as well. The foremost among such events was the 1990 adoption by the Indian government of structural adjustment reforms. Influenced by the World Bank–International Monetary Fund combine, structural adjustment has meant the gradual withdrawal of state patronage for higher education and a coterminous privatization of that sphere. However, the government itself is confused and has been dithering about the policy to be adopted in this area.

State investment in education in India has always been inadequate for meeting the needs of "education for all." An international comparison revealed that on a list of 86 countries, India (with an expenditure of 3.8 percent of the gross national product on education) ranked only 32nd in terms of public expenditure on education as a proportion of GNP (Shariff and Ghosh, 2000, p. 1396). Structural adjustment has meant a drastic cut in public expenditure on higher education: Between 1989–1990 and 1994–1995, the share of higher education in public expenditure decreased from 12.6 percent to 6 percent (Tilak, 1996). The annual growth rate of public expenditure on university and higher education has also declined (Shariff and Ghosh, 2000, p. 1400).

Thus, the state, which had hitherto been the dominant partner in funding higher education, is finding it increasingly difficult even to maintain the same level of funding for higher education. There is no gainsaying that financial constraint does not affect all sectors of higher education equally.[1] Invariably, nonprofessional courses are more adversely affected than their professional counterparts (Varghese, 2000, p. 22). Furthermore, efforts to privatize higher education by encouraging private agencies to set up institutions of higher learning have enjoyed limited success in general education and nonprofessional courses. Thus, state universities and their affiliated colleges (which in the early 1990s accounted for a stupendous 88 percent of over four million student enrollments in higher education) are the ones in the financial doldrums. It is at these state universities and affiliated colleges that the bulk of the teaching community is employed.

The gradual withdrawal of the state from higher education has been accompanied by its inability to address the need for reforms within conventional higher education. The National Policy on Education (Government of India, 1985) and the Program of Action (Government of India, 1986), and their review by the Acharya Ramamurti Committee (Singh, 1991) all preceded the structural adjustment reforms. Unanticipated were both the phenomenal fall in the demand for conventional courses in the B.A. and B.Sc. streams and the remarkable spurt in the demand for courses in such areas as computer science and information technology, biotechnology, and management studies.

The fact that the state is unwilling and unable to invest in the new areas of education explains the growth of private institutions offering professional courses (Tilak, 1999). The so-called self-financing institutions are of uneven quality, ranging from centers of excellence to roadside teaching shops. These new entrants to the arena of higher education raise questions of autonomy and accountability (since they do not depend on state funding), as well as issues concerning teachers (like qualifications and recruitment, career options, pay, and working conditions). While these institutions are recognized by the state governments and universities, they do not submit themselves for assessment and accreditation.[2] Thus, private initiatives in higher education are fraught with serious implications for the academic profession.

Similarly, the falling demand for conventional courses and their drop in enrollment have impacted the academic profession. The lack of a link between conventional programs and the job market seems to have become all too apparent to students and their parents.[3] The situation has become aggravated with the structural adjustment reforms, which demand types of knowledge and skills generally not possessed by conventional degree holders. It is only natural that many of those (including the "not-so-bright" ones) who have hitherto been using the conventional programs as waiting rooms are either seeking early entry into the job market, with the option of obtaining formal university qualifications later, or entering programs that carry better job prospects. Those who still seek out the conventional graduate programs are generally the leftovers and dregs or first-generation students from rural and indigent backgrounds, especially those who are supported with financial assistance by the government. All this has implications for the size, working conditions, and quality of the academic profession.

THE DECLINING STATUS OF TEACHERS

During the five decades of independence (since 1947), India has built up a massive system of higher education (Government of India, 1999). In 1998–1999, there were 214 universities (198 state government established and 16 central government established), 38 institutions "deemed-to-be universities," 11 institutes of national importance, 9,703 colleges, and 887 polytechnics. The system now employs 321,000 teachers and caters to 6,755,000 students—though estimated to be covering barely 6 percent of the population in the relative (17–23) age group.

An important consequence of this phenomenal unbridled expansion, which took place mainly in the 1970s and 1980s, was the unprecedented demand for teachers. An increasing number of postgraduates churned out by the state university system found in teaching an easy employment avenue. The cumulatively adverse consequences of the reckless manner in which teachers were recruited and allowed to function soon became evident. In its all-India survey covering a "proportionate stratified random" sample of 8,450 teachers (2,144 in 27 universities; 6,306 in 304 colleges), 2,114 students, and 1,658 members of the wider community, the National Commission on Teachers (NCT) recorded the "widespread feeling that no profession has suffered such downgrading as the teaching profession." This finding was reflected in "the low esteem given to the profession and the unfavorable image of teachers held by parents, students and by the people at large." What is more pathetic, "even members of the teaching community have a low esteem of their own profession" (NCT, 1985, p. 21).

The NCT's observations referred to the situation in 1983–1985, and there is no evidence suggesting that the situation has changed for the better since then. During the author's discussions with teachers in various parts of the country, these observations were reiterated and confirmed: A sense of helplessness, an attitude of indifference, and a feeling of cynicism are shared by a majority of teachers. In what follows we shall discuss some dimensions of the erosion in the social status of teachers.

Shrinking Academia

That there has been a 16-fold increase in the number of teachers in higher education in five decades, and that it now stands at 321,000, seems to be indicative neither of the healthy growth of the academic profession nor of its strength and vitality. Not only have the prospects of employment as a full-time college or university teacher become dim, but the security of employment, once taken for granted in the academic profession, is becoming increasingly problematic. Even in 1983, the NCT found that only 70.7 percent of university teachers and 68.5 percent of college teachers had permanent employment with all statutory benefits. The others were either "temporary" (with no guarantee of continuation) or "ad hoc" (appointed against a leave vacancy for a short period of time) lecturers. Besides, new categories of teachers such as "part-time" lecturers (who teach for a specified number of teaching hours in a week) and "guest" lecturers (who help the college or department "to complete portions of the syllabus") have been added. Such teachers are paid on an hourly basis, and they do not enjoy other privileges that go with a permanent or even a temporary or an ad hoc teacher.

The decline in the employment prospects in the academic profession has to do with the combined effect of the structural adjustment reforms and the market forces operating in higher education. The expansion of conventional programs seems to have outstripped the demand for them by students. For a few years now, many colleges have been experiencing a decline in enrollments in these programs, some of which are in dire straits with absolutely no enrollment. For some grant-in-aid private colleges—government-aided, privately managed colleges—it has even become difficult to assemble workloads for their teachers. In fact, there are cases of colleges in which the teachers, to protect their positions, pay for the dummy admission of students. Reputed colleges are not in the red yet, but it may be only a question of time for them, too.

Most state governments have imposed an embargo on the recruitment of teachers. This has meant a freeze on the establishment of state-supported colleges, the downsizing of the number of permanent teachers at existing colleges, and optimization of resources by redeployment of teachers through a policy of transfers. Besides, most governments have also introduced "voluntary retirement schemes" (giving incentives to teachers to retire early), and some state governments are contemplating reducing the retirement age for college and university teachers.

For instance, in the southern state of Karnataka, the government has shut down 27 colleges run under its aegis and redeployed the surplus teachers. The state government has imposed a virtual ban on further recruitment of teachers in grant-in-aid colleges. According to the Karnataka state minister for higher education, in October 2000 there were 692 vacancies in 152 government first-grade colleges in the state. There has been no recruitment in these colleges for nearly eight years now. After the reallocation of excess staff in grant-in-aid colleges to these government colleges, the vacancies had come down to 364. The decision to delink the first two years of postsecondary education—referred to as the "plus-2" stage—from colleges and assign it to higher secondary schools will further downsize the employment opportunities in colleges.

Not surprisingly, temporary part-time teaching positions have almost become a permanent feature of higher education in Karnataka, and in some other states, too. The opening of private colleges indiscriminately and without meeting the primary requirements has resulted in a spurt of part-time teachers. With the adoption of the University Grants Commission (UGC) norms for the appointment of teachers since 1985 and the abolition of the practice of appointing temporary teachers in 1992, these part-time teachers have been consigned to the margins of the profession. In 1996, there were about 2,500 part-time teachers in the state, most of whom were between 25 and 35 years of age. Those over the age of 30 who have been part-time teachers for over ten years hardly have suitable opportunities for alternative employment.

Periodically, part-time teachers with ten or more years of service have, on humanitarian grounds, brought political pressure to bear on the state government to "regularize" their appointments. Courts of law have also been sympathetic to their cause. To overcome the administrative and financial problems of such "backdoor entry" into the academic profession encouraged by grant-in-aid private colleges, in June 1996 the Karnataka government directed all colleges in the state to abolish the post of part-time lecturers and reappoint retired lecturers on a contract or hourly basis.

Significantly, in the northern state of Bihar, there has been no regular recruitment of lecturers since 1981. In 1984, following the decision made at a conference of vice-chancellors, the system of appointing ad hoc lecturers was officially introduced. In December 1989, the Supreme Court of India directed the Bihar government to regularize the services of ad hoc lecturers who were in service as of February 1989. By 1991, the number of such ad hoc teachers had risen to 1,400, and more than 50 percent of them had doctoral qualifications.

The downsizing of the academic profession through a freeze on recruitments, redeployment of excess staff, appointment of guest lecturers, and other measures is now a pan-Indian phenomenon. Moreover, it is not confined to the conventional liberal science colleges. The Swaminathan Committee (AICTE, 1994), which examined the issue of resource mobilization in technical education, strongly advocated the rationalization of teaching workload so as to reduce the salary bill. It recommended the reduction of the share of regular faculty to 60 percent and the appointment of the remaining 40 percent on a part-time and contract basis. It also suggested the downward revision of the staff-to-student ratio from 1:10 to 1:15 in degree courses and from 1:11 to 1:20 in diploma courses. However, in burgeoning fields such as computer science, information technology, and biotechnology, in which the expansion has been most rapid, there is a dearth of qualified teachers. This is notwithstanding the fact that the new pay scales have made the job of college teacher more attractive.

The shortage of teachers is most acute in medical education. For instance, in the western state of Maharashtra, the 11 state-run medical colleges are struggling with a perennial staff shortage: 40 percent of 150 posts of professors, 30 percent of the 600 posts of associate professors, and 60 percent of the 750 posts of lecturers remain vacant. Furthermore, an estimated 15 percent of current teaching staff are expected to leave, while an estimated 18 percent are seeking voluntary retirement. In the case of medical education, other considerations inhibit qualified persons taking up teaching jobs. At the end of long years of study, the doctors holding M.B.B.S. and M.S./M.D. qualifications get the same salary as those who have a M.A./M.Sc. and Ph.D. in the sciences and the humanities. Also, the service rules in the government colleges are archaic and despotic, and there are restrictions on private practice.

Be that as it may, the bulk of the teaching community is engaged in general education. The employment opportunities here have almost dried up, and those seeking entry into the profession are those who have been lingering on a part-time or ad hoc basis. That existing teachers are finding it difficult to get adequate workloads does not augur well for the academic profession. The state policy of downsizing the profession is likely to have an adverse impact on the already low morale and commitment of teachers.

Skewed Composition and Parochialism

The academic profession was traditionally a preserve of the higher social strata, and this was matched by the high prestige attached to it. However, rapid expansion of higher education has altered the social profile of the profession. In its all-India survey, the NCT found that most teachers in higher education (83 percent in the universities and 77 percent in the colleges) were male and that they came from "educated families." As compared to their male counterparts, female teachers came from "families in economically better off occupations." Interestingly, the educational level of teachers serving in private, unaided colleges was relatively better than that of teachers in other types of colleges (NCT, 1985, pp. 12–13), as the unaided colleges have greater autonomy in the matter of recruitment of teachers and they are also free from governmental norms favoring indigent caste groups and communities.

Analyzing secondary data, Chanana has found that "[t]he higher the level of education, fewer are the women teachers...." In 1993–1994, women constituted about 35 percent of teachers at the higher secondary level (i.e., those who teach classes 9–12), whereas they formed only 18.8 percent of the 291,048 teachers in higher education. Their representation was better in affiliated colleges (21.0 percent) as compared to university colleges and teaching departments (11.6 percent). Furthermore, as regards academic leadership roles in higher education, women still lag behind: "Very few women occupy positions of authority and decision-making in the universities and colleges. However, rotation of headship of departments has helped bring women in the forefront as academic leaders" (Chanana, 2000, p. 1019).

That the proportion of women in the academic profession fell during the 1990s is intriguing. Perhaps the downsizing of the profession is negatively affecting the entry into it of women much more than of men. Some state governments have introduced a provision of reservation of teaching posts for women. Moreover, women's colleges generally prefer women candidates for teaching jobs. Addressing the problem of career development of women teachers, the NCT proposed granting age relaxation to women whose careers are interrupted due to problems of maternity and child care needs, and counting total experience (not continuous service) for benefits of vertical mobility (NCT, 1985, p. 51).

Interestingly, the decline of the academic profession has coincided with the broad basing of its social composition. Thanks to the policy of protective discrimination, during the last three decades, a significant section of the candidates belonging to the scheduled castes, scheduled tribes, and other "backward classes" (categories referring to the traditionally disadvantaged sections of the population identified for special benefits and concessions) have entered the academic profession. In some states teaching posts are reserved for specific caste groups and communities, and the provision governing the de-reservation of posts has been scrapped. The new entrants into the profession, many of them being the first generation in their caste and community groups to have acquired postgraduate qualifications, have hardly any exposure to the cultural moorings of the profession and are confused about the ethos of a profession in decline.

What is more disconcerting, parochialism and inbreeding have become an integral part of higher education. Educational institutions run by minority religious communities have always shown preference for candidates belonging to their own religion or sect, and similarly, those dominated by particular caste groups have shown bias in favor of their caste fellows. Universities and state governments, too, prefer candidates belonging to their own jurisdictional areas or states. "The sons or daughters of the soil" policy, which has been in tacit operation since the mid-1970s, is being vigorously followed now. The adoption of the state language as the medium of instruction in higher education precludes the vast majority of eligible candidates from outside a given state.

A committee that reviewed the workings of central universities in the early 1980s found "the extent of inbreeding in one of these at 85 percent in the case of first-level appointments and at 92 percent in the case of faculty members coming from the states concerned and the adjacent states" (NCT, 1985, p. 42). The tendency for such inbreeding, which must have strengthened during the last two decades, can hardly be different in state universities and affiliated colleges. The NCT's recommendation that "at least 25 percent of teachers recruited at the initial level should be from outside the state in which they are recruited" (NCT, 1985, p. 43) has not found favor with the UGC, nor will it be acceptable to the state governments.

As a matter of policy, the UGC is supportive of interinstitutional mobility of teachers, which, it is hoped, will infuse fresh blood into a system that would otherwise become stagnant. It is also expected to result in cross-fertilization with varied experiences at different institutions. While inbreeding inhibits mobility, it is not the only impediment in the path of mobility. A reduced chance of promotion as a result of moving from one institution to another would discourage lecturers and readers from leaving the institution at which they work. State government rules governing the recognition of services of teachers for retirement benefits being rigid, senior teachers would be wary of moving out of the state in which they work.

Voluntary mobility of teachers should not be confused with their compulsory transfer. Transfer of teachers from one government college to another has been in vogue right from the beginning. Some states have now incorporated a provision of transfer of teachers from one grant-in-aid private college or university to another. This provision is generally invoked for administrative convenience or for the redeployment of excess teaching staff. However, there have been cases in which transfer has been used as an instrument of favor by powerful politicians or as a mode of punishment for "recalcitrant" teachers. Both inbreeding and compulsory transfer are deleterious to the system of education: While inbreeding results in systemic stagnation, compulsory transfer erodes teachers' sense of loyalty and commitment to the institution.

Deficient Professional Preparation

Studies on college teachers have invariably underlined the sad deficiency of academic preparation of the people entering the profession and their declining commitment to it (Bali, 1986; Bhoite, 1987; and Kaul, 1993, pp. 224–240). The NCT (1985) also bemoaned the fact that most of the teachers were making a living and not following a vocation. This has, no doubt, a lot to do with the deplorable standards obtaining at the postgraduate level of education. More important, however, is the fact that for decades most master's degree holders easily found employment at colleges, and even at universities, with absolutely no training in or orientation to teaching, and with doubtful aptitude for that vocation.[4]

It is to arrest this trend and to ensure proficiency in the subject and aptitude for teaching or research on the part of candidates aspiring to become teachers that the UGC introduced the scheme of the National Eligibility Test (NET). This test, jointly conducted by the UGC and the Council for Scientific and Industrial Research (CSIR) at about 90 centers in the country, is held twice a year. Many state governments have been permitted by the UGC to conduct a State Level Eligibility Test (SLET), which is treated as equivalent to the NET. In some states, to cope with various demands, the standard of the SLET has been so appallingly diluted and the norms so brazenly flouted that the UGC has had to withdraw the permission granted to some states to conduct the SLET.

While implementing the latest pay scales in April 1998, the UGC had made it "optional for the University to exempt Ph.D. holders from NET or to require NET, in their case, as a desirable or essential qualification for appointment as lecturers...." (UGC, 1998, p. 2). Candidates who had completed their M.Phil. degree or had submitted their Ph.D. thesis in the concerned subject by 31 December 1993 were exempted from the NET requirement. However, in April 2000, the UGC withdrew this provision and made the NET mandatory even for Ph.D. holders. This has been criticized by some educationists as devaluating the Ph.D. degree and discouraging serious research (Seethi, 2000). However, others have welcomed it as an effort to enforce basic standards in the recruitment of lecturers at universities and colleges. The latter argue that the latest pay scale package contains substantial incentives to Ph.D. degree holders, both at the entry point into the profession and because of the career advancement within it (Krishnakumar, 2000). The Supreme Court of India has upheld the UGC's directive, making the NET a necessary qualification for recruitment of lecturers at universities and colleges.

The idea of an "all-India qualifying test" at the stage of entry into the academic profession was first mooted by the NCT. Such a test inter alia was expected to eliminate "the evils of nepotism, parochialism and undesirable pressures and interventions" in the recruitment of teachers. More important, it would arrest "the backdoor entry of less competent people into the profession through ad hoc and temporary appointments" (NCT, 1985, p. 43). As a screening mechanism, the NET is a step in the right direction. Nevertheless, the lacuna of inadequate academic preparation of teachers for discharging their professional responsibilities remains. Professions such as architecture, law, and medicine require their prospective recruits to undergo a specified period of internship. Even a high school teacher is expected to acquire the bachelor of education (B.Ed.) degree. To become a lecturer at a college or university, however, absolutely no prior training or experience is necessary.

While this anomaly is recognized, there has never been a consensus among educationists as to the nature and extent of the additional qualifications required for entry into the academic profession. Insistence on a research degree (Ph.D. or M.Phil.) has become counterproductive: The rush for enrollment in doctoral programs, following the UGC's decision in the 1970s to make a Ph.D. the minimum qualification, has downgraded the quality of doctoral research in universities. Whether the acquisition of a Ph.D. degree necessarily enhances the performance of a teacher, especially at the first-degree level, remains a moot question. It appears that the UGC has not been able to resolve the conceptual confusion centering around the differences between "quality," "qualification," and "minimum eligibility," and their relative importance.

Ritualization of the Training Program

Important as the question of qualification and screening at the point of entry into the profession is, the need for postinduction training and periodical professional enhancement can hardly be exaggerated. The NCT observed that every person inducted into the academic profession must undergo "a training course relating to a proper orientation toward the profession and its values, skills in pedagogy, curriculum construction, use of audiovisual aids, communication skills, educational psychology, evaluation methods, as well as the use of the medium of instruction" (NCT, 1985, p. 47). This was reiterated by the National Policy of Education (GOI, 1986), and following the recommendation thereof, in 1987 the UGC introduced a permanent and structured program known as Academic Staff Orientation Scheme (ASOS).

Based on a review of past experience with some variants of the B.Ed. program (a minimum requirement for employment as a high school teacher), the UGC was clear that the structure and objectives of ASOS had to be different. Starting in 1987, the UGC has established at least one Academic Staff College (ASC) in each state with the mandate to improve standards of teaching through "orientation courses" (focusing on pedagogy and the social relevance of education, for young lecturers) and "refresher courses" (providing up-to-date information on the contents of various disciplines, for senior lecturers). By 1994, 45 ASCs had been established, and 71,385 teachers had attended the orientation (27,675) and refresher (43,710) courses (Chalam, 1994, p. 43). While in terms of sheer numbers, the ASOS appears to be impressive, the functioning of the ASCs is far from satisfactory. As Indiresan has observed,

> Often the[ir] implementation is unplanned, and the approach is ad hoc particularly in the appointment of directors, in selecting resource persons and in conducting programs. There has been very little monitoring. The overview of the functioning of Academic Staff Colleges indicates that while some of them are functioning well, mainly due to the competence and enthusiasm of the director, about a third have been observed to be decidedly poor. (1993, p. 317)

Resource crunch is the familiar refrain heard from many directors for the poor quality courses organized by their ASCs. There is an element of truth in this: The scheme is fully financed by the UGC, and often its grants are not made available to the ASCs on time. Furthermore, to save on cost, various restrictions are imposed, thereby curtailing the freedom of course coordinators in matters of selection of resource persons and pedagogic practices. Considering that pay scales of teachers have been amply revised upwards, the UGC should consider making the teacher-participants pay for the orientation and refresher courses, as they are materially benefiting from these courses. This would enable the UGC to impose its standards and inject an element of competition among the ASCs to improve the quality of their courses.

The ASCs were entrusted with conducting programs for properly orienting the new entrants into the profession, and improving the knowledge and skills of those already in it. To instill a sense of seriousness, an element of compulsion has also been introduced: Those entering the profession are required to attend an "orientation course" before they complete their probation. Those in service are required to attend two "refresher courses" to become eligible for career advancement or promotion. As with all initiatives carrying an element of compulsion, the original objectives behind the establishment of ASCs have been lost and the courses ritualized.

Since the recruitment to permanent posts at universities and colleges has dwindled, there is not much of a rush for orientation courses. However, the case of refresher courses is different, as the number of teachers seeking career advancement is very large and the facilities at the ASCs are meager. To meet the demand for such courses the UGC has been providing grants to university departments without ASCs to organize refresher courses. Besides these, groups of teachers under the banner of their subject associations and in collaboration with their respective university departments are organizing "self-financed" refresher courses.

Most refresher courses, irrespective of whether they are organized by the ASCs or university departments (UGC-sponsored or self-financed), are conducted as a formality and generally lack the advanced academic orientation expected of them. Teachers attend such courses out of compulsion. Even the academically oriented among them complain about the poor quality of the courses. The routine assessment of the courses by the teacher participants is unreliable, as this is often fabricated by course coordinators and ASC directors. An independent review of the working of ASCs and the courses organized by them is overdue.

The Job: Retailing Stale Knowledge

Besides ASOS, the UGC has initiated the College Science Improvement Program and the College Social Science and Humanities Improvement Program to enhance the quality of teachers. Permanent teachers desirous of acquiring doctoral qualification are given paid leave for two or three years under the Faculty Improvement Program. Teachers interested in pursuing research are offered grants for projects. Financial assistance is extended to teachers to attend seminars, symposia, and workshops. Promising young teachers with a research proclivity are offered funds under the Career Award Scheme, and the renowned among senior teachers are given National Associateship grants.

Human resource development schemes such as the above, which are far more widely available now, were expected to improve the quality of teaching over the past several decades and thereby benefit students. However, they do not seem to have yielded the expected results. Most teachers do not avail themselves of the opportunities for professional development.[5] Even those teachers who have participated in the Faculty Improvement Program or similar facilities have at best obtained only a research degree, but they have not taken their advanced training into the classroom. Some teachers in the program are reported to have spent the time on activities other than research. Similar complaints have been raised about the provision of sabbatical leave extended to university teachers. Not surprisingly, the UGC has become more restrictive in awarding these Faculty Improvement fellowships.

As far as other facilities for research are concerned, they are made use of by university teachers. Even so, the research output of teachers is quite low. In the mid-1980s, the NCT found that one-third of university teachers had not published any articles and about three-fourths no books. Hardly 25 percent of college teachers had published any articles, and only 10 percent any books. While the lecturers had the fewest publications to their credit, a considerable percentage of readers and professors, too, had not published anything at all. Also, very few teachers (less than 10 percent at colleges and less than 20 percent at universities) are engaged in supervising research. Overall, the activities of university teachers were directly related to their positions, but in terms of research their performance was no different from that of college teachers (NCT, 1985, p. 48). The situation does not seem to have improved over the past two decades.

Even with teachers who are engaged in research, seldom is there a link between laboratory or field work and what is taught in the classroom. This is mainly due to the fact that the main function of the teacher is to prepare the students to complete the annual examination successfully. There has hardly been any innovation in curriculum and pedagogy in higher education, and the traditional lecture (monologue) method combined with the dictation of notes is prevalent in most undergraduate programs (in colleges) and in many postgraduate programs (in universities). Thus, the undue emphasis on certification rather than on the teaching and learning process—a classic case of the proverbial tail wagging the dog—has distorted the orientation of higher education and the academic profession.

There are complaints that even this conventional function is not performed properly by teachers. In its survey the NCT found "that 22 percent of college and 18 percent of university teachers are perceived by students to take their classes without preparation. (In one state the figure is 38 percent.)" Eighty-six percent of college teachers and 90 percent of university teachers did not come up very well even in the perception of members of the wider community (NCT, 1985, p. 38). The fact that this has not been a matter of concern for teachers unions is reflective of the endogenous decay of the profession.

SALARY, CAREER, AND SERVICE CONDITIONS

While professional development and teaching performance have seldom been the concern of teachers or their associations, the issues of salary, career, and service conditions have always been high on their agenda. In fact, teachers have often blamed inadequate salaries and unattractive service conditions for the deterioration in the status of the academic profession. Members of the wider society also rate the profession very poorly with respect to material benefits and perquisites.

Revised Pay Scales

There have been two major revisions of pay scales during the last two decades—first, in 1987 (following the Mehrotra Committee Report) and the latest one, in 1998 (following the Rastogi Committee Report). Both these revisions were preceded and followed by agitations and negotiations. The dust raised by the 1998 revision has not yet settled. However, it can hardly be denied that in absolute, if not relative, terms the current pay scales are the best deal that the teachers have had, considering the nature and quantum of work that they do and the little accountability that is demanded of them.

The pay package set forth by the UGC in 1998 was generally in effect from January 1996 (UGC, 1998). In addition to the basic pay, teachers are entitled to dearness allowance (linked to the cost of living index according to a fixed formula, currently forming 41 percent of the basic pay), city compensatory allowance (in the case of those working in cities), and subsidized housing or house rent allowance. The details of the pay scales for different categories of teachers and the gross salary drawn by them at the start and end of the scale, as of March 2001, are shown in table 8.1. Besides, the teachers are entitled

Table 8.1   Current pay structure of university and college teachers

Category	Gross salary (start)	Gross salary (end)	Service span
Lecturer	Rs. 12,600 (U.S.$271)	Rs. 21,180 (U.S.$455)	20 years
Lecturer (senior scale)	Rs. 15,720 (U.S.$338)	Rs. 23,832 (U.S.$513)	16 years
Lecturer (selection grade) or reader	Rs. 18,840 (U.S.$405)	Rs. 28,668 (U.S.$616)	15 years
Professor	Rs. 25,704 (U.S.$553)	Rs. 35,064 (U.S.$754)	13 years

Note: The gross salary (in Indian rupees) drawn by the different categories of teachers is computed from the pay scales notified by the UGC (1998), and the rates of dearness allowance (41 percent of the basic pay), house rent allowance (15 percent of the basic pay), and city compensatory allowance (Rs. 120) in vogue in March 2001. Figures in parentheses refer to equivalents in U.S. dollars, at Rs. 46.50 per dollar in March 2001.

to leave or home travel concession and medical aid. Schemes such as provident fund (lump-sum retirement benefit) and pension are available to teachers, and retiring teachers get a gratuity.

Parity with select categories of government employees has always been a crucial issue whenever pay scales of teachers are revised. Obviously, teachers do not compare themselves with doctors, architects, chartered accountants, and advocates, as these professionals have independent "practices" for which there is both private and institutional demand. The only avenue of private practice for teachers—namely, engaging tuition classes—is regarded as unethical if they are employed in permanent positions. In recent negotiations, too, teachers' organizations demanded the observance of the principle of parity with Class "A" civil service. However, a careful analysis of the Rastogi Committee Report, on which the new pay scales are structured, reveals that teachers have at best marginal parity with Central Services Group B officers (Kumar and Raina, 1997).

While the UGC pay package has been accepted in principle all over the country, there are significant variations in its implementation by different states. This was partly due to the authority vested in the state governments to alter pay scales and to schedule their implementation. Citing its inability to bear the financial burden, the Bihar government has flatly refused their implementation. While some states have postponed the date of implementation, a few have not given arrears accruing from delayed implementation of the scales. Also, the UGC scales are applicable to both university and college teachers, and most states have denied the new pay scales to junior college teachers. Although the UGC notification made reference to part-time teachers, no state has implemented the relative provisions, and these teachers continue to get a paltry sum of Rs. 1,200 (U.S.$26) per month.

Thus, the gross salary of different categories of teachers in terms of their institutional affiliation varies across the country. Even so, the increased gross salary of teachers has brought practically every teacher into the income tax net. Teachers now pay income tax ranging from 10 to 30 percent of their gross salary, minus payments toward medical insurance premiums, contribution to charitable funds, and a certain proportion of the amount spent on house rent. They are also entitled to a tax rebate for certain investments and savings. There is a perception among teachers that whatever benefits they receive through the new pay scales will soon be neutralized by inflation and income tax. Even so, with their current salaries teachers can lead comfortable middle-class lives.

The central government met 80 percent of the additional expenditure to be incurred by the state governments for implementing the revised pay scales. However, the state governments were to take over the entire liability in April 2000. This massive burden is now being felt. In many states teachers' salaries are not being paid regularly. The Maharashtra government has frozen the rate of dearness allowance. The Karnataka government is holding in abeyance the proposed 15 percent cut in grant-in-aid to private colleges. In other words, though the new pay scales have been introduced, the teachers cannot take the security of their salaries for granted.

Career Advancement

The academic profession has traditionally been pyramidal, with more lecturers than readers, and more readers than professors. This has meant that, irrespective of academic achievements and proven professional development, after a specified span of service in any cadre a teacher was destined to stagnate. While stagnation for a good proportion of their working span was inevitable for all categories of teachers, the period of stagnation was relatively longer for lecturers. The NCT survey found that "a fairly large percentage of teachers stagnated in the same position for ten or more years." The degree of stagnation was "alarming" at the level of lectureship. Stagnation had a negative impact for college teachers more than for their university counterparts: "[O]nly 13 percent of college teachers were promoted at least once as compared to 45 percent in the case of university teachers" (NCT, 1985, p. 47).

If stagnation has an adverse impact on the morale and commitment of the members of a profession, career advancement makes the profession attractive to qualified and capable young persons. Earlier efforts at introducing career advancement, such as "merit promotion schemes" and "time-bound personal promotion schemes" were inspired by the bureaucratic notion of promotions. Laying emphasis on seniority, with little or no reference to professional achievements, these practices resulted in reckless promotion of teachers to the levels of readers and professors. For this reason, the NCT recommended making career advancement and professional development "contingent upon each other and intertwined in a sequential system" (NCT, 1985, p. 47).

To give adequate and suitable opportunities for vertical mobility to teachers at multiple stages in their careers, in addition to the new pay package the UGC has incorporated a career advancement scheme based on the professional development of teachers. A person entering the system as a lecturer can move into the grade of lecturer (senior scale) after four, five, or six years, depending upon whether she or he possesses a Ph.D., a M.Phil., or only a master's degree. After five years of service as lecturer (senior scale), a teacher can move into the grade of reader (if he or she holds a Ph.D.) or lecturer (selection grade) if he or she does not hold a Ph.D. That is, only teachers with a Ph.D. are eligible to become readers. After eight years of service as a reader, a teacher will be eligible to be considered for appointment as a professor. In brief, a person with a Ph.D. degree entering the system as lecturer can hope to become a reader after 9 years of service and a professor after 17 years of service. The UGC has also proposed the creation of a new position—namely, "professors of eminence" for teachers who have completed 28 years of service.

Unlike "promotion," career advancement, theoretically at least, is not automatic. The cases of teachers seeking career advancement are processed through selection committees that are generally constituted for direct selection of candidates. Consistently satisfactory performance appraisal reports are a prerequisite for advancement from one level to the next. Participation in orientation and refresher courses, publication of books and articles, attendance at seminars and conferences, a good record in teaching, contribution to educational innovation and curriculum development, enhancement of the corporate life of the institution, engagement in extension and field outreach activities, and so on, are given due credit. While this scheme is well thought out, its effective implementation cannot be taken for granted.

For instance, hitherto performance has not been a criterion for recognizing and rewarding teachers. While some states (for instance, Karnataka) have instituted annual awards for "best teacher" in different faculties, these make but little impact as motivators or incentives to improve teachers' performance. It is true that good performance by the students in the examination may bring some credit (and satisfaction, too) to teachers, either directly or indirectly. However, poor student performance is seldom used as a measure to admonish or punish teachers. Incidentally, except in some universities and the Indian Institutes of Technology and Management, peer review or student evaluation of teachers is virtually nonexistent in the traditional system of college and university departments. Any proposal for such a review or evaluation would be vehemently opposed by teachers' unions.

The National Policy on Education envisaged the creation of an open, participative, and data-based system of teacher evaluation. It even contemplated laying down "norms of accountability ... with incentives for good performance and disincentives for nonperformance" (GOI, 1986, p. 25). Following this, the UGC announced a format and procedure for "self-appraisal" by teachers, both at the time of entry into the profession and annually after that. However, this has either not been introduced or perfunctorily done, and as such it has seldom formed the basis of any action. Realizing this, the UGC has now made "consistently satisfactory performance appraisal reports" mandatory for career advancement.

Retirement Age

Presuming that a college or university teacher enters the profession between age 23 and 27, depending upon whether a Ph.D. degree has been acquired or not, and the prevailing retirement age being 58 to 60 years, he or she will be able to put in a maximum of 31 to 37 years of service. This is at least 4 to 5 years less than in other jobs in the government service. This curtails the retirement benefits available to teachers, and is also a loss to society, as the teachers with experience and professional achievement will no longer be actively involved in education. Accordingly, a marginally higher retirement age for teachers has invariably been recommended by various committees and commissions.

The UGC guidelines regarding the pay package fixed the retirement age for university and college teachers at 62 years. It further recommended that a maximum of three years' benefit in service should be provided to teachers entering the profession with a Ph.D. degree, so that they, too, will get full retirement benefits accruing to those who have put in 33 years of service. While the UGC is categorical that "no extension in service should be given," it allows universities and colleges to re-employ a superannuated teacher up to the age of 65 years, according to certain prescribed guidelines (UGC, 1998, pp. 14 and 18).

However, only the central universities have accepted the recommendation to fix the retirement age at 62. State governments have retained the existing retirement age, as they fear that it would lead to agitations by government employees demanding similar upward revision of the retirement age. Considering the growing unemployment among educated persons, it would be indefensible for any state government to raise the retirement age. More important, at a time when state governments have meager resources for higher education and are consciously pursuing a policy of downsizing the number of teachers—even offering voluntary retirement schemes—raising the retirement age would be unthinkable.

As a consequence, there is no uniformity in retirement age across the country. For instance, in Kerala, college teachers retire at age 55; in Goa, they retire at 60; in Karnataka, while university teachers retire at 60, college teachers retire at 58; and in central universities, teachers retire at 62. The disparity in retirement age has definitely agitated teachers, but they have not succeeded in convincing the state governments in their favor. Arguing that the retirement age of 62 was part of a package and that state governments are bound to implement the package in toto, some teachers have approached courts of law.

Workload

The new pay package also prescribed the number of teaching days and the workload of teachers (UGC, 1998, pp. 13–14). A minimum of 180 "actual teaching days" in a year has been stipulated for universities and colleges. Universities are to devote 72 days and colleges 60 days for admissions formalities and the evaluation of students. University teachers are entitled to eight weeks vacation, and in lieu of two weeks vacation they are to be credited with one-third of the period of "earned leave." College teachers are entitled to 10 weeks vacation and no earned leave, unless they are asked to work during the vacation, for which they, too, will be credited with one-third of the period as earned leave.

The workload of teachers in full employment has been fixed at not less than 40 hours a week for 30 working weeks (180 teaching days) in an academic year. Of these, a workload of 16 hours of direct teaching per week has been prescribed for lecturers (including senior scale and selection grade lecturers), and 14 hours per week for readers and professors. Professors engaged in administration, research, and extension are given a remission of two hours. Furthermore, it is necessary for teachers to be available for at least five hours daily at the university or college.

Though not wholeheartedly, these prescriptions have been accepted by teachers. However, going by past experience, they are sure to be observed more in breach than in practice. For instance, university and college calendars have formally incorporated the prescribed number of "working weeks" and "teaching days" and the stipulated duration of vacations. However, in connivance with their teachers' unions, some universities have cleverly introduced midterm holidays, called a "break" rather than "vacation." Also, whether the prescribed number of teaching days is adhered to is not audited by any agency. It is common knowledge that with admissions taking place late, innumerable official and unofficial holidays, and strikes by students, teachers, and nonteaching staff, the loss of working days is quite high. Not surprisingly, a study conducted by the National

Institute of Educational Planning and Administration (New Delhi) on the work ethos in colleges found that the average number of working days to be as low as 87 (Indiresan, 1993, p. 313).

During the last two pay revisions, teachers in various parts of the country have gone on prolonged strikes. During the 1990s, teachers at Delhi University established a record of sorts by going out on strike for 110 days and 90 days. On both these occasions, settlements were reached between their association and the UGC, with the teachers promising to take extra classes to complete the syllabus. This record was broken by Bihar university and college teachers in 2000 when they went on a 131-day strike, demanding the implementation of the new pay scales. To regularize the teaching schedule, the universities in Bihar pretended to make 2000–2001 a year sans holidays. The soft attitude of the government toward the unionization of teachers is largely responsible for this state of affairs.

As regards workload, since it is helpful in downsizing the number of teachers, state governments have insisted that teachers in grant-in-aid colleges have the prescribed workload (especially the number of direct teaching hours). They have even redeployed the excess teaching staff (i.e., those having less than the prescribed minimum workload). Though universities have not as yet enforced the minimum workload, fearing downsizing of staff, some university departments have inflated actual workloads.

Whether the quantum of "direct teaching hours" (16 hours per week for lecturers and 14 hours per week for readers and professors) is a pedagogically sound norm is a different question, which does not seem to bother either teachers or the UGC. Of concern is teacher truancy and absenteeism. In many colleges and university departments, other than regularly engaging the classes allotted to them, the teachers are not even available in the institution for the stipulated five hours a day. Attempts by concerned vice-chancellors, principals, and heads of departments to police teachers have not yielded the desired results. In Kerala state, the proposal to authorize police officials to inspect the colleges to ascertain the presence of teachers during working hours has provoked strong reactions from teachers.

It must be conceded that the everyday work of college or university teachers does not lend itself easily to policing or surveillance, as in the case of many other kinds of work. As sociologist Béteille observes, "The only guarantee of regular and dependable service in their case is the internal censor of each teacher. It is this internal censor that is damaged and sometimes destroyed by frequent and prolonged strikes" (2000, pp. 86–87). It was only in 2000 that the Delhi

University Teachers' Association formally recognized the problem of teacher absenteeism and decided to monitor its members for performance.

Academic Autonomy and Professional Ethics

Academic freedom is defined as "the freedom of the academic community to deal with academic matters the way they like, but in keeping with their professional obligation" (Agrawala, 1994, p. 216). Barring a few rare exceptions, this freedom can hardly be said to have been denied to teachers in India. In fact, instances of teachers abusing it abound, and these take many forms, such as nonperformance of role obligations (teaching and research), resisting change in curriculum and pedagogy, indulging in malpractice in evaluation, and so on. This calls for governmental intervention, just as it raises the question of the professional obligation of teachers.

Governmental Intervention

Teachers often complain about encroachment on academic autonomy by governmental authorities. In view of the state funding of higher education, governmental intervention in university and college affairs is inevitable and necessary. Most universities cannot generate their own funds, or not to a sufficient extent, and the predicament of private colleges receiving grant-in-aid by the government is no better. If a university or college does not comply with government guidelines, it is reined in through holding back grants. Complaints of corruption and inefficiency on the part of universities and colleges make such intervention justifiable. There is no denying that a political stranglehold over education in the guise of governmental control has been on the increase.

In a rare case, when the Maharashtra government tried to encroach on the autonomy of Kholapur University, the teaching community of that university fully backed its vice-chancellor's stand. Politicians in Maharashtra, as in most other states, have business interests in private educational institutions. They found the vice-chancellor to be uncooperative—by insisting on the observance of stipulated norms. Since the existing law did not enable them to remove the vice-chancellor, they began lobbying to amend the law itself (Morkhandikar, 2000).

While the teaching community's defense of university autonomy in the above case was justifiable and laudable, its critique of governmental intervention per se is not. For instance, in Karnataka, teachers have been opposing as retrograde the bill to amend the Karnataka State Universities Act, which among other things envisions making the assessment of teachers a precondition for their promotion, and the evaluation by students a part of teaching activities.

Interestingly enough, teachers have been opposing the idea of granting autonomous status to select colleges. Against a target of 500 colleges envisioned by the UGC, barely 110 have received autonomous status, even ten years after the target year (1990). Teachers know that autonomy calls for greater levels of accountability than currently exist. They are also worried about the enormous control that autonomous status will give the college management over them. Thus, if greater accountability is demanded of them and if their job security is at stake, teachers are ready to part with academic autonomy.

Code of Professional Ethics

Academic freedom bestowed upon teachers is premised upon the societal responsibility that their profession carries. Their job involves the dispassionate search for truth and its transmission and application as recorded knowledge. They are a model to their clients—namely, the students—who are highly impressionable. As such, compared to other groups in society, a higher level of conduct and obligation is expected from teachers (Agrawala, 1994, p. 215), and due deference is shown to them.

The NCT devoted considerable attention to the professional ethics and values of the academic community. It examined some of the basic areas of concern in which the profession has failed miserably. Its survey covered not only the values that university and college teachers cherish but also those they ought to uphold, but do not. The NCT urged teachers to "scrutinize their own value system" and "impose severe restrictions on themselves" so as to raise the profession to "the highest moral level" and earn "the most honorable place" in society (NCT, 1985, pp. 55–65). The National Policy on Education (UGO, 1986) also recognized the need for introducing discipline into the system. However, no plan of action has followed in this regard.

Ideally, as in any profession, the enforcement of a code of ethics among teachers should be undertaken by their professional organizations. The All India Federation of University and College Teachers Organizations (AIFUCTO), the only voluntary professional association of university and college teachers at the all-India level, does not exercise any authority, moral or otherwise, over the members of the profession as regards their professional conduct. Of course, the AIFUCTO does not deny the importance of a code of professional ethics for teachers. As early as 1976, it even adopted a "note" in this regard, though it made the observance of this code contingent on the creation of congenial conditions and incorporation of a section on the rights of teachers (UGC, 1986, Appendix 3).

In 1988, a task force set up by the UGC prepared a Code of Professional Ethics, in consultation with AIFUCTO (UGC, 1989). There was nothing in this code that could make it enforceable, either legally or morally. Interestingly, it was addressed to the vice-chancellors rather than the teachers. The AIFUCTO did not think it necessary to call on its members to observe this code even as a moral obligation (Agrawala, 1994, p. 220). Apparently, most universities have ignored this code, though the UGC has been ritually reminding them about its implementation. Even those universities that have implemented this code in principle are not enforcing it in practice, either due to administrative inertia or the fear of agitation by teachers.

Given the inability of universities to enforce such a code, there have been cases of high courts imposing financial liability on universities in the form of damages for negligence arising from acts of omission or commission by their teaching or nonteaching staff. Some universities have also been directed to take action against erring employees (Agrawala, 1992). Damages have also been awarded against universities under the Consumer Protection Action 1986 for acts of negligence by teachers and nonteaching employees. It is a sad commentary on the profession that instead of evolving its own mechanism for accountability, the universities and colleges are compelled to act on the directions of courts.

Private Tuition

On the subject of professional ethics, one issue that has attracted the critical attention of governmental authorities and members of the public alike is that of the charging of private tuition by college teachers. The rise of "shadow education" conducted through "coaching classes" is closely related to the falling standards of formal education. With existing colleges being unable to teach effectively and students wanting to sharpen their competitive edge, private tuition has become a vital supplement and is thriving.[6] The dynamics of this dimension of education is seldom covered in discussions on the privatization of higher education (Tilak, 1999).

Since teachers involved in coaching classes are, by and large, formally employed by colleges on a full-time tenured basis, private tuition raises the question of professional ethics. Their being engaged in private tuition is a reflection of the substandard teaching that their colleagues in the college are doing. Yet, since so many students go in for private tuition, teachers do not take their teaching in the college seriously. Often the success of private tuition is attributed to the "leaking" of exam questions by such teachers. In brief, private tuition seems to have caught teachers and students in a vicious circle.

Private tuition, given by individual teachers or by a group of teachers (coaching classes), is not a new phenomenon. It has now been institutionalized as a money-spinning enterprise. Institutes offering coaching classes even advertise in newspapers and claim credit for the success of students in the merit lists of various examinations. Some well-reputed teachers have taken voluntary retirement or resigned from their jobs at their colleges to engage in this profitable enterprise.

The UGC has always been critical of college and university teachers engaging in private tuition but has not been able to do anything about it. State governments have been ambivalent about private tuition: While in principle they are opposed to it, many a state has introduced special coaching classes for students belonging to the scheduled castes and tribes and other "backward classes." Some states have issued administrative orders banning private tuition and coaching classes, but find it impossible to implement the ban. Raids by income tax authorities on the houses of some private tutors have not deterred teachers from engaging in private tuition. Teachers' organizations are silent on the whole issue.

THE POLITICIZATION OF ACADEMIA

After a prolonged period of political apathy, mainly due to its middle-class moorings, the teaching community has been gradually politicized since the dark era of the Internal Emergency (1975–1977), when the rights guaranteed by the Constitution were suspended to meet the political crisis confronted by the then Congress Party–led government. That this politicization has coincided with the decline of the profession is a matter of concern. It is not that academia has become an arena of party politics or ideological battles, though in some universities even this has occurred. Rather, "the politics of scarcity" has a more direct bearing on the academic profession now than ever before, and teachers have become politically more conscious, though their organizations have been unable to articulate this change in attitude effectively.

Democratizing Academia

Over the last two decades, measures of internal democracy have been introduced in the university system, though not uniformly. Most universities have created "departmental councils" consisting of all or some teachers for arriving at decisions affecting the routine functioning of departments. More important, in many universities "headship" of the department is no longer a permanent office: Under the principle of "rotation of headship," the position of the head or chairperson of the department is occupied in turn by teachers of specified designations for two to three years. In the most liberal system, all permanent teachers, irrespective of their designation (professor, reader, or lecturer), head the department by rotation. In the most conservative system, rotation is among professors, if there is more than one professor; and if there is no professor in the department, then among readers, if there is more than one reader; or else the sole professor or reader will continue to hold headship until a second professor or reader joins the department. Between these two extremes, there are many variants.

Apart from instilling a sense of democracy into the functioning of departments, rotation of headship envisages that a person by becoming the head of the department does not have to invest all her or his time in routine administration and committee work. In practice, however, rotation of headship has generally implied a discount on leadership. With the duration being limited, even serious academics are shy of taking initiatives. At universities where even a lecturer can become the head of the department, this has upset the conventional notions of academic hierarchy. Furthermore, personal animosities among teachers assure that no head will be effective.

Some universities with a large number of affiliated colleges have bifurcated the boards of studies, so that each subject now has two boards, one each for the undergraduate and postgraduate courses. Election of teachers' representatives on academic councils, senates, and executive councils has been in vogue for many decades now. This principle is now being extended to other bodies (such as, boards of studies) and offices (such as deanships). For example, in 1994 the Maharashtra government introduced the constitution of boards of studies through election among teachers. Since "elections are often based on considerations other than academic [considerations]...they may result in choosing representatives of limited appeal and sometimes in monopolies of groups with a set of opinions" (NCT, 1985, p. 37). But then, if academics cannot make democracy work, who else will?

Teacher Unionism

Practically every university has one or more teachers' unions, euphemistically called organizations or associations, to distinguish themselves from working-class trade unions. The growth in the number of such unions does not necessarily denote a healthy development of the profession. An integral feature of the development of teachers' unions has been the multiplicity and fragmentation of organizations. Generally, two significant divisions within the ranks appear along geographic and professional lines. To this split may be added the process of further segmentation within any one level. For instance, college teachers in every state are organized at state and local levels, as postgraduate and undergraduate teachers or both, as employees of government or private or university-managed colleges, and as members of particular faculties. Such a proliferation of teachers' unions through a process of fragmentation and segmentation has weakened the teachers' movement.

Studies on the unionization of teachers have revealed that it "does not necessarily ensure their collegial participation or promote professionalization among them." By and large, teachers' unions have steered clear of larger political issues. "This neutrality is both a cause and a consequence of the unions being controlled by professional teachers and not professional union organizers." The number of professional union organizers is larger than it was two decades back, and they will determine the future of teacher unionism (Jayaram, 1992, pp. 161–162). Anticipating the grave consequences of this trend, the UGC and state governments have repeatedly emphasized the need to depoliticize academia.

Teachers' unions are no longer so strong. Even the AIFUCTO does not command the mass support it once did. Given teachers' middle-class preoccupation with economic issues, pay scales are the only issue on which they can be mobilized. On closer review, it appears that whatever strength teachers' unions manifest is not due to any intrinsic qualities, but rather to the soft attitude of the government toward them. It is amazing that even when teachers go on prolonged strikes, the principle of "no work, no pay" is not applied to them. It was only during the Internal Emergency (1975–1977) that teacher unionism was suppressed and their leaders jailed. But that period was the nadir of civil liberties in the country anyway, when the darkness of the eclipse of law spared none (Jayaram, 1992, p. 167).

The predominant mode of protest of teachers' unions continues to be the strike. Generally, the issues that sustain any strike have to do with pay scales, promotions, and service conditions. In the recent past teachers in various parts of the country have gone on strike demanding the implementation of new pay scales and modifications of some of their provisions. The Andhra Pradesh Federation of University Teachers' Associations protested against the increasing privatization of higher education and government intervention in the administration of universities. The proposed cut in grant-in-aid to private colleges has also been protested by some teachers' unions.

The pattern of agitation by teachers is by now well established—consisting, among other things, of protest rallies, hunger strikes, marching to the chief or education minister's house, absenteeism from work, and finally, the boycott of examination work. Given the government's response to strikes by much stronger unions of employees in the telecommunications, postal, insurance, and banking sectors during the last few years, teachers cannot take the material success of their strikes for granted. Even state-level agitations are running out of steam. After their record-breaking 131-day strike demanding the implementation of new pay scales, the teachers in Bihar had to return empty handed. An agitation led by a political party-backed teachers union (e.g., the West Bengal College and University Teachers' Association) will be a different story as long as the party (the Communist Party of India-Marxist) is in power. However, as Béteille has observed, "A strike by university teachers cannot be judged solely by its material successes or failure. Far more important in the long run are its moral and psychological effects. The typical pattern of such strikes is to create an initial state of exhilaration which is then followed almost inevitably by a state of profound demoralization" (2000, p. 85). Repeated strikes have not only affected the self-esteem of teachers, but have also caused them to become increasingly less inhibited about being absent from work. As noted earlier, a major cause of deterioration in higher education is teacher truancy, and this has not only given teachers a bad name but also spoiled the public image of the academic profession.

CONCLUSION

Higher education in India is undergoing an uncertain transition. With the structural adjustment reforms and liberalization of the economy, the state is gradually shedding its responsibility for higher education.

The UGC has been virtually reduced to being a mere funds-disbursing agency, incapable of enforcing its own recommendations. Educationally, the Indian university system has progressively become nominalized and marginalized. Being outside the purview of the UGC and largely of the state governments as well, the emerging private educational initiatives may have a different story to tell, but for that we will have to wait.

As regards the academic profession, the decline that was noticed over a quarter century ago is now almost complete. Entering the profession with no prior professional preparation other than a postgraduate degree, assured of tenure, doing unchallenging work without any accountability, and with their performance being no more than its own reward, teachers at colleges and universities have been largely reduced to the lowest common denominator. Every laudable policy to improve the situation has been ritualized in practice. It is true that the situation is better at some centers of excellence, the institutions of national importance, and a few university departments and colleges. They are, however, drops in an ocean of mediocrity.

Ironically, the improvements in pay scales and service conditions have come at a time when the profession has almost sunk to its lowest level. Teachers are largely happy with the pay package, but they are also worried about the gradual withdrawal of state patronage to higher education. In the meantime, politicians and people are agreed, and not unjustifiably so, that teachers are a pampered lot who are getting more than they deserve.

NOTES

Author's Note. These data and cases are obtained from a review of the literature; personal discussion with a cross-section of teachers, administrators, and students in and from several states; personal experience of working in two state universities and doing committee work for the University Grants Commission; and observations during visits to various universities and colleges on official work. I thank Ms. S. Suma for her assistance in the collection of data and all those who provided valuable information and insights on the problems and prospects of the academic profession in India.

1. The term "higher education" suggests too much homogeneity. Even a casual review of the educational landscape would reveal the enormous structural diversity within the system of higher education in India (Jayaram, 1997, pp. 77–79). There are different types of universities: the central, the state, the deemed (those recognized by the UGC although not established as such by an act of Parliament or a state legislative assembly), and the

open; Institutions of National Importance; an assortment of institutes under the umbrella of the Councils of Research in Science and Industry, Social Sciences, History, and Philosophy; and the traditional institutions of higher learning in religion and theology; besides a large number of varied university or government-run colleges and grant-in-aid or purely privately run affiliated colleges. These institutions vary as to their objectives and sources of finance, and the academic preparations, abilities, motivations, and commitment of their faculty and students. The focus of this chapter is mainly on the state universities and their affiliated colleges, which together employ the overwhelming majority (over 85 percent) of teachers.

2. As a step in the direction of quality control in higher education, following the National Policy of Education (GOI, 1986), in 1994 the UGC set up an autonomous body called the National Assessment and Accreditation Council (NAAC). Initially the scheme of assessment and accreditation was voluntary, but the idea of an external institution doing this was not well received even by the universities and government-aided colleges. By the end of May 2001, the NAAC had been able to assess and accredit only 173 institutions. Now the scheme has been made mandatory for universities and grant-in-aid colleges but remains voluntary for the self-financing private institutions.

3. Being aware of the disorientation of the conventional programs, the UGC recommended the introduction of job-oriented courses at the first degree level. However, there does not seem to be a proper understanding of the failure of a similar program introduced at the level of two-year postsecondary education. Many universities have introduced a job-orientation component in their undergraduate curriculum, mainly to avail themselves of the funds provided by the UGC for the purpose, and as such the program has been ritualized.

4. The NCT (1985, pp. 13 and 49) found that nearly 73 percent of college teachers had only a master's degree in the subject, and 6.6 percent a M.Phil. degree. Even among university teachers only a little over 61 percent possessed the doctoral degree, and 32.6 percent had nothing but a master's degree. Most teachers with a doctoral degree (76.5 percent of college and 77 percent of university teachers) acquired it after joining the profession.

5. Professional development among teachers is pathetically low: about two-thirds to three-fourths of college teachers, and about one-third to half of university teachers have never participated in any seminar, summer school, workshop, or research project (NCT, 1985, p. 49). If any college teachers are now attending orientation and refresher courses under ASOS, it is because their career advancement is dependent on it.

6. More college teachers than university teachers are engaged in private tuition, and it is in greater demand for science and mathematics courses and courses in the English language.

REFERENCES

Agrawala, S. K. (1992). Courts and accountability of universities: Some recent developments. *University News*, 30, 19–24.

———. (1994). Code of professional ethics: Accountability of teachers. In M. V. Mathur, Ramesh K. Arora, and Meena Sogani (Eds.), *Indian university system: Revitalization and reform* (pp. 214–227). New Delhi: Wiley Eastern.

All India Council of Technical Education (AICTE). (1994). *Report of the high power committee for mobilization of additional resources for technical education*. New Delhi: AICTE.

Bali, A. P. (1986). *College teachers: Challenges and responses*. New Delhi: Northern Book Centre.

Béteille, André. (2000). *Chronicles of our time*. New Delhi: Penguin Books India.

Bhoite, U. B. (1987). *Sociology of Indian intellectuals*. Jaipur: Rawat.

Chalam, K. S. (1994). *Performance of academic staff colleges in India*. Visakhapatnam, India: Andhra University Press and Publications.

Chanana, Karuna. (2000). Treading the hallowed halls: Women in higher education in India. *Economic and Political Weekly*, 35, 1012–1022.

Chitnis, Suma, and Altbach, Philip G. (Eds.). (1979). *The Indian academic profession: Crisis and change in the teaching community*. New Delhi: Macmillan.

Government of India (GOI). (1985). *Challenge of education: A policy perspective*. New Delhi: Ministry of Education.

———. (1986). *Program of action: National policy on education*. New Delhi: Ministry of Human Resource Development.

———. (1999). *India 1999: A reference annual*. New Delhi: Publications Division, Ministry of Information and Broadcasting.

Indiresan, Jayalakshmi. (1993). Quest for quality: Interventions versus impact. In Suma Chitnis and Philip G. Altbach (Eds.), *Higher education reform in India: Experience and perspectives* (pp. 309–333). New Delhi: Sage.

Jayaram, N. (1992). India. In Bruce S. Cooper (Ed.), *Labor relations in education: An international perspective* (pp. 157–169). Westport, CT: Greenwood.

———. (1997). India. In G. A. Postiglione and G. C. L. Mak (Eds.), *Asian higher education: An international handbook and reference guide* (pp. 75–91). Westport, CT: Greenwood.

Kaul, Rekha. (1993). *Caste, class and education: Politics of the capitation fee phenomenon in Karnataka*. New Delhi: Sage.

Krishnakumar, G. (2000). Case for national eligibility test for teachers. *Economic and Political Weekly*, 35, 2369.

Kumar, T. Ravi, and Raina, Badri. (1997). Rastogi committee pay structure: Disincentives reinforced. *Economic and Political Weekly*, 32, 1985–1990.

Morkhandikar, R. S. (2000). Politics of education in Maharashtra. *Economic and Political Weekly,* 35, 1991–1993.

National Commission on Teachers (NCT). (1985). *Report of the national commission on teachers, II, 1983-85.* New Delhi: Controller of Publications.

Seethi, K. M. (2000). UGC's disincentives for Ph.D. *Economic and Political Weekly,* 35, 1895–1896.

Shariff, Abusaleh, and P. K. Ghosh. (2000). Indian education scene and the public gap. *Economic and Political Weekly,* 35, 1396–1406.

Singh, Amrik. (1991). Ramamurti report on education in retrospect. *Economic and Political Weekly,* 26, 1605–1613.

Tilak, Jandhyala B. G. (1996). Higher education under structural adjustment. *Journal of Indian school of Political Economy,* 8, 266–293.

———. (1999). Emerging trends in evolving public policies in India. In Philip G. Altbach (Ed.), *Private Prometheus: Private higher education and development in the 21st century* (pp. 127–153). Chestnut Hill, MA: Center for International Higher Education, Boston College.

University Grants Commission (UGC). (1986). *Report of the committee on revision of pay scales of teachers in universities and colleges.* New Delhi: UGC.

———. (1989). *Report of the task force on code of professional ethics for university and college teachers.* New Delhi: UGC.

———. (1998). *UGC notification on revision of pay scales, minimum qualifications for appointment of teachers in universities and colleges, and other measures for the maintenance of standards.* New Delhi: UGC.

Varghese, N. V. (2000). Reforming education financing. *Seminar,* 494, 20–25.

9

THE ACADEMIC WORKPLACE IN
PUBLIC ARAB GULF UNIVERSITIES

André Elias Mazawi

This chapter examines the professoriate and the organization of the academic workplace in public universities currently operating in the six Gulf Cooperation Council (GCC) member states—Bahrain, Kuwait, Oman, Qatar, Saudi Arabia, and the United Arab Emirates (UAE). The academic workplace and the university professor, as they are known in the West, represent recent phenomena in Arab societies, particularly in the Gulf states. The emergence of the contemporary Arab university (*jami'a*)—as well as the notion of the Islamic university (Al-Assad, 1996, pp. 26–30)—and of the role of the university professor (*ustaz*) represent cultural products associated with the Western colonial encounter with Arab societies and the expansion of Western higher education systems (Rida, 1998, pp. 119–120). In this sense, the contemporary Arab academic workplace differs from earlier institutional models of knowledge formulation and transmission as they were historically and socially experienced in Arab and Muslim societies (Makdisi, 1981; Berkey, 1992; see particularly the critique of Al-Assad, 1996).

The contemporary academic workplace, as an organizational context and space of practice, introduced new paradigms of knowledge into Arab societies (Rida, 1998). By the same token, it triggered new political realities and power struggles, gender and class based. By sanctioning what expertise and knowledge are, the academic workplace emerged as a new center of authority—one that confronted or interacted with established bases of power, on which the established elites often relied for legitimacy. By implication, the university campus (*al-haram al-jami'i*) and the academic department (*daïra*)—as spaces where new forms of competence and knowledge are located and administered—have become crossroads where constellations of power were being negotiated and practiced and where the pursuit of knowledge is legitimized—but in the same vein, closely monitored and scrutinized.

OVERVIEW

The Expansion

In the relatively short period of time since the 1960s, Gulf higher education developed into a diverse system of specialized organizations. Currently, the region is experiencing another period of rapid expansion of higher education. In the UAE, for example, four universities (two of them American) and one private polytechnic were opened within just a two-year period (1997–1998). By 2000, 18 universities were in operation (public and private), in addition to well over 80 colleges and teacher training institutes. Additional higher education venues are being explored as well. Calls for distance higher education and the development of university outreach programs are increasingly being heard and debated. In all GCC member states, motions were initiated, and laws passed, albeit not without opposition in some cases, for the purpose of legalizing the creation, accreditation, and operation of private or community universities and colleges. Within less than three decades, from the 1970s onward, Gulf societies were gradually, yet effectively, incorporated into Western credentialing systems. An American university professor, acting as consultant to the UAE Ministry of Higher Education, described the academic model underlying the vision of one of the country's new universities (a women's university) as follows: "[The University] is being designed to reflect the typical design of colleges and universities in the U.S. so it will qualify for accreditation—or its equivalent—by regional and professional accrediting associations in the U.S. That, in sum, will facilitate transfers to U.S. institutions and entrance to US graduate programs for students with those aspirations" (Halloran, 2000, pp. 329–330).

Some observers have described the expansion of Gulf higher education as part of the nation-building policies of governments eager to enhance regime legitimacy and foster a sense of nationhood among their citizenry ('Abdallah, 1994, p. 128; Al-Ebraheem and Stevens, 1980, p. 205). Other factors singled out have included political and military rivalries among Gulf governments (Al-Misnad, 1985), as well as the discovery of extensive oil reserves—the world's largest—in the Gulf states (Shaw, 1997). The oil industry, which came to constitute the major component of Gulf state revenues, brought much wealth and many resources. These processes triggered radical and complex transformations of Gulf political, occupational, social, and cultural structures. The "oil boom" of the 1973–1982 period and the ensuing growth—in oil production potentials, economic returns, and state apparatuses—enabled local governments to embark on the provision of large-scale free social services in many spheres of life, including education and higher education (Gilbar, 1997).

Yet, as the expansion of higher education initially occurred in societies with particularly low literacy rates, it necessitated an overwhelming and ever-growing reliance on imported multinational labor. Currently, "nonnational" labor, as it is called, constitutes around 70 percent of the entire GCC workforce (Al-Sulayti, 2000, p. 275) and between one-quarter and two-thirds of the total population, depending upon the state concerned. Nonnationals are recruited mostly from countries located in North Africa, the Eastern Mediterranean, and Southwest and Southeast Asia. To a lesser extent, the recruitment of higher professional and scientific cadres also occurs in Western countries, particularly among naturalized Arabs.

The excessive reliance on multinational labor prompted Gulf governments to undertake, though with limited success, policies aimed at "nationalizing" employment—that is, increasing the representation of nationals in various economic sectors by affecting hiring practices. Nationalizing the teaching force and university faculty staffs was designated as one of the priorities. However, graduates' career patterns, as well as state patronage practices, are still based on the importance of academic credentials as privileged channels for entering senior civil service positions. As a result, Gulf policymakers often complain that credentials obtained by citizenship-holding graduates are largely unrelated to job requirements and economic productivity, expressing a "growing gap between quality of graduates and labor market needs" (Mograby, 2000, p. 299).

Governance and Administration

Gulf public universities vary considerably, not only in size but also in their histories. (See table 9.1.) Saudi Arabia has the largest and oldest higher education institutions in the Gulf. With eight universities founded between 1957 and 1998, Saudi universities employ almost 75 percent of all Gulf public university faculty members. The university systems of Kuwait, the UAE, and Qatar date back to the 1960s–1970s, while Oman and Bahrain have universities dating from the mid-1980s. The first university in Bahrain was founded in 1980

Table 9.1 Public Arab Gulf universities, by state

State	University	Foundation	Students	Faculty
Bahrain	Arabian Gulf University	1982	600	66
	University of Bahrain	1986	8,000	322
Kuwait	Kuwait University	1966	18,000	929
Oman	Sultan Qaboos University	1986	6,000	409
			(c. 2/3 women)	
Saudi	Al-Imam Muhammad bin			
Arabia	Saud Islamic University	1974	30,000	1,648
	Islamic University of			
	Al-Madinah Al-Munawwarah	1961	5,032	216
	King Abdel-Aziz University	1967	39,000	2,143
	King Fahd University for	1963	7,000	693
	Petroleum and Minerals		(men only)	
	King Saud University	1957	32,000	2,224
	King Faisal University	1975	10,969	295
	Umm Al-Qura University	1981	20,600	1,103
	King Khalid University	1998	*not available*	*not available*
Qatar	University of Qatar	1973	9,000	432
UAE	United Arab Emirates	1977	17,000	626
	University		(70% women)	

Note: Student figures are rough approximations derived from official and nonofficial sources. Statistics may vary considerably between sources. Figures for Saudi Arabian universities were displayed on the Saudi Ministry of Information Website at: www.saudinf.com. The figures for faculty are number of teaching staff as coded for the present paper. For Al-Imam Muhammad bin Saud Islamic University, figures exclude faculty teaching in the university's branches located outside the Kingdom of Saudi Arabia.

as a GCC regional graduate institution, operated through the Arab Bureau of Education for the Gulf States.

Gulf universities also differ in terms of their organizational characteristics. Most Gulf universities offer *Shari'a* (Islamic jurisprudence) and Islamic studies alongside other disciplinary fields, though in distinct faculties or departments. Five specialized universities, three of which operate in Saudi Arabia, command attention. Saudi Arabia has three universities devoted to Islamic studies and jurisprudence: the Islamic University of Al-Madinah Al-Munawwarah (Medina), Al-Imam Mohammad bin Saud Islamic University (Riyadh), and Umm Al-Qura University (Mecca). The first also enrolls Muslim students from abroad, and the second operates branches in Djibouti, Indonesia, Japan, Mauritania, the UAE, and the United States. Al-Imam Mohammad bin Saud is currently the world's largest Islamic university. Also in Saudi Arabia, the Dhahran-based King Fahd University for

Petroleum and Minerals specializes in science and technology. The UAE-based Ajman University of Science and Technology maintains undergraduate faculties of business administration, computer sciences, engineering, education, pharmacy, and health and a doctoral program in dentistry. The Bahrain-based Arabian Gulf University operates a regional graduate institution with two colleges: the College of Medicine and Medical Sciences and the College of Post Graduate Studies in the areas of technology and education.

During the primary phase of expansion, from the late 1960s onward, Egyptian and other Arab faculty and experts played a dominant role in extending the Egyptian higher education model—itself patterned along eighteenth- and nineteenth-century French and British academic traditions—into nascent Gulf universities. However, with the adoption of the course unit system starting from the mid-1970s, the Egyptian legacy—particularly in university governance and administration—was significantly eroded (Al-Ebraheem and Stevens, 1980, pp. 208–209; Safi, 1986, p. 423).

As state institutions, Gulf public universities are directly managed by the state, through a ministry of higher education (Oman, Saudi Arabia, UAE), a ministry of education (Bahrain), or a combined ministry of education and higher education (Kuwait, Qatar). No buffers operate between the state and higher education institutions, such as an agency like a University Grants Commission that exists in some other countries. Where a council of higher education is provided for, as in Kuwait and Saudi Arabia, it operates as part of the state bureaucracy and is headed by a state minister. Council membership comprises representatives of relevant state ministries, senior civil servants, and university rectors. This body approves new higher education institutions, units, and programs and appoints university rectors and vice rectors. It also defines the necessary laws and regulations pertaining to the governance and administration of universities and other institutions of higher education.

In turn, the university council is headed by the state minister. The minister appoints faculty deans and supervises rectors and university operations. Within each university, the rector serves as deputy chairman of the university council (*Majles al-Jami'a*), which is made up of vice-rectors, deans, and several external members. This body is entrusted with all university matters, including those pertaining to the granting of academic credentials, the approval of textbooks and curricula of existing departments, and the formulation of recommendations to be forwarded to the relevant bodies.

Women are overwhelmingly represented in Gulf universities and colleges, amounting in some states to between one half and two-thirds of the total student population. Men, by contrast, continue to seek higher education opportunities abroad since career advancement is now increasingly linked to foreign postgraduate study, given the continuing low prestige of Gulf universities (Mahdi, 1997, p. 23).

Organizationally, Gulf universities operate under centralized administrations, in which hierarchical relations of authority affect faculty participation. For instance, at Kuwait University, it has been found that deans were perceived by faculty members as exerting the most significant influence on the day-to-day administration (Al-Khalifa, 1990). The same finding was reported with respect to Saudi Arabia. Deans exert a decisive effect on the evaluation of departmental chairpersons (Al-Karni, 1995, pp. 39 and 45). Deans' attitudes toward the selection of department chairpersons were found to be motivated also by considerations of reelection. Deans also did not trust outside sources for the evaluation of department chairpersons, perhaps "reflect[ing] the practice of centralizing authority in the dean's hands" (Al-Karni, 1995, pp. 45, 55, 56). UAE University has been described as having a highly centralized governance system, with major decisions being made at the top ('Abdallah, 1994). The broader social and political culture seems to have pervaded the university space and affected administrative and governance styles.

Some writers have attempted to explain the causes underlying the rigid administrative system in Gulf universities. The high annual expansion rates, in the number of students and faculty members, have been cited as placing great pressure on university governance and administration (Al-Saadi, 1997, p. 198). It has been pointed out that Gulf universities are primarily perceived as apparatuses facilitating the implementation of state development policies, thus exerting mounting pressures on the university's internal organization (Alghafis, 1992; 'Abdallah, 1994, pp. 127–128; El-Shibiny, 1997, p. 169). This link with national development has led to an "inflation" in the bureaucratic structure of Gulf universities—"gradually expanding towards academic departments, exhausting faculty members in daily and routine paper work disconnected from direct academic and research interests" ('Abdallah, 1994, p. 219).

Gender

Gender segregation in Gulf higher education presents contrasting realities. In Saudi Arabia, gender segregation is the norm and the law. Gender-specific faculties operate in tandem, for instance at King

Abdel-Aziz University. Other Saudi universities enroll men only, as in the case of King Fahd University for Petroleum and Minerals (Zehery, 1997, p. 20). In some cases, arrangements are provided for *intisab*, a method that enables women to study without necessarily attending classes (Al Rawaf and Simmons, 1992; El-Sanabary, 1992, p. 11). Other arrangements enable women to follow lectures in parallel classrooms via closed-circuit television (Al-Saadat and Afifi, 1990).

In some of the smaller Gulf states men and women may share the same facilities, for instance at Kuwait University. The latter has been operating de facto as a coeducational institution, in spite of opposition-sponsored legislation to the contrary in 1996, much objected to by the government. Similarly, in Oman in the Departments of Home Economics and Geography, at Sultan Qaboos University, explicit attempts have been undertaken to create coeducational programs of study (Al-Qaydi, 1999, p. 584; Moosa, 1999).

The Chancellor of the University of Sharjah (UAE), Isam H. Zabalawi, recently recognized that, being "a non-coeducational institution," the university he presided over was nonetheless "aware of the patriarchal tilt in the society." He further acknowledged that "evidently, we have a long way to go to achieve GE [gender equity] in the University, especially in the direction of hiring more female faculty members, appointing more females in top administrative positions, adding gender studies to our programs and setting up a graduate program in women studies" (Zabalawi, 2000).

Gulf Private Higher Education

The issue of private higher education lies beyond the scope of this chapter. Nonetheless, it bears observing that private institutions have become an option, though still confined to the smaller Gulf states. For instance, in the UAE, two American universities, at Sharjah and Dubai, started operating in the late 1990s. Also in the UAE, American, British, Australian, and lately Canadian universities are competing fiercely for students, undergraduate as well as graduate. Some competitors have judged it cost-effective to open locally operated undergraduate programs, if not full-fledged universities (Caeser, 1999).

In Oman, the latest GCC member state to operate a national university, the issue of private higher education has been debated in recently organized conferences and forums. A private university venture, Sohar University, reportedly started operation in September 2001. Named for and located in Oman's second-largest city, the university, owned by the "for-profit" Oman Education and Training Investment Groups, has links with the University of Queensland in Australia. It will offer undergraduate programs, leading to degrees in engineering, business management, and information technology (Del Castillo, 2001). Buraimi University College, an Omani higher education joint venture between the Nuaimi Group and California State University, was expected to begin operations in September 2002. This college will offer bachelor's degrees in business management, financial management, data systems management, computer sciences, accounting, marketing, and the English language (Buraimi Varsity, 2001).

In June 2001, plans for a private venture were announced in Qatar, with the sponsorship of Cornell University and the Qatar Foundation for Education, Science and Community Development. The venture expects to establish, by 2004, the Weill Cornell Medical College in Qatar, which "will offer a complete medical education in Qatar leading to a Cornell University M.D. degree, based on the same admission standards and curriculum as the New York campus" (Qatar Foundation, 2001).

In Kuwait, a law on private universities was recently passed by the National Assembly. Before the law passed, a clause was inserted by an Islamic-oriented bloc—to the government's displeasure—prescribing gender segregation as a condition for their operation (Kuwait Islamists, 2000). In Saudi Arabia, where no foreign or private universities operate, regulations enabling the establishment of what are termed "community colleges" have been established.

Private higher education is fee paying. Where it operates, it tends to serve relatively more established socioeconomic groups. Curricula tend to be patterned after Western conceptions of liberal education. They further seek to enable graduates to undertake advanced studies abroad (Caeser, 1999).

ORGANIZATION OF THE ACADEMIC WORKPLACE

GCC member states have attempted, through the Arab Bureau of Education for the Gulf States, to promote a coordinated policy on university equivalence, faculty recruitment, and terms of employment (Al-'Areed, 1993). Additional initiatives were undertaken by the Association of Arab Universities, to which Gulf universities are also affiliated. While Gulf universities share broadly similar recruitment and employment policies, they differ nonetheless on others. It should be noted, however, that the differential status of "national" versus "nonnational" faculty members constitutes a basic divide replicated in Gulf universities.

Recruitment, Terms of Employment, and Salary

Faculty members who are nationals are appointed to tenure-track positions. To promote their representation, special incentives are provided—including scholarships for studying abroad and for completing doctoral and other professional degrees. Academic returnees are also given priority in terms of their recruitment into tenure-track faculty positions. By contrast, the terms of employment for nonnational faculty members, who are recruited internationally, are regulated by contract. Nonnationals are usually recruited when specific departments or programs lack nationals with the necessary qualifications or expertise.

Faculty members who are nationals are government employees. As such, they are subject to civil service laws and regulations. In Saudi Arabia, for example, in terms of compensations, emoluments, and honors, faculty members are treated like other state employees, based on the following formula: instructor, 8th rank; lecturer, 9th rank; assistant professor, 12th rank; associate professor, 13th rank; and professor, 14th rank (Council for Higher Education, 1996). Retirement age for national faculty members is set at 60 years according to the Islamic calendar. Under certain conditions it can be extended to age 65.

The terms of employment of nonnational faculty members are covered by different bylaws. Nevertheless, nonnationals are also subject to the laws governing the civil service, in terms of expected behaviors. In Saudi Arabia, for example, the *Regulations for the Employment of Non-Saudis in the Universities* govern the recruitment of nonnationals (Council for Higher Education, n.d.). When recruited, non-Saudi faculty are employed on a contractual basis "for a period of one year or less, renewable for the same period or for the period set by the university" (art. 6). Among other obligations, a nonnational faculty member "is committed to follow the regulations and instructions in force in the kingdom, and he and his dependents must respect the traditions and customs prevalent in the kingdom, avoid offending religion or interfering in politics" (art. 45).

In the UAE, a contract nonnational faculty member is first subject to a two-year probationary period. Once tenure is granted, the contract is extended for a period of four years, renewable for an additional period of three years (*Faculty Handbook,* 2000). Thus, UAE nonnational faculty members—who constitute the majority of faculty members at UAE University, remain insecure in their positions with severe repercussions for their professional freedom and academic productivity ('Abdallah, 1994, p. 131).

Income is tax free in the Gulf states. Moreover, in all Gulf states, salaries include allowances, premiums, and fringe benefits over and above the basic salary granted to both nationals and nonnationals. For instance, senior university administrators, faculty deans, department heads, and center directors receive a monthly allocation on top of their salaries.[1]

In the case of nonnationals, the university further provides faculty members with housing and furniture, or assumes the related costs. Under specific conditions, the university may also assume the expenses related to the education and travel for a nonnational's dependents. Health services are similar to those enjoyed by nationals. A given number of airplane tickets to and from the host country are covered by the university.

Separate salary scales exist for nonnationals. Yet, the salary of nonnationals varies greatly (in some cases, informally) based on the individual's citizenship, credentials, and field of expertise. Regulations often justify a substantial deviation, of up to 100 percent, from the officially prescribed payment scales, with respect to "teaching faculty members who have rare specializations, or those enjoying scientific reputation or expertise or high skills, or who have excellent qualifications obtained in one of the famous universities, including physicians" (Council for Higher Education, n.d., art. 9). Nonnationals—particularly those from Western nations and associated with leading universities—have a greater opportunity to negotiate salaries beyond the officially specified scale. Salaries of nonnationals are reportedly 40 to 50 percent higher than those of nationals (Tansel and Kazemi, 2000, p. 89). In the same vein, nonnationals from less established countries may have less leverage in this respect.

Promotion

Nationals receive tenure when they are appointed (Al-Ebraheem and Stevens, 1980, p. 217). Promotion requests are initiated at the level of the department in which the faculty member is employed. (For a comprehensive comparative study of Gulf faculty promotion procedures, see Al-'Areed, 1993.) Table 9.2 lists the minimal requirements for promotion to the ranks of associate and full professor in Saudi Arabian universities, as examples. (See Appendix for a detailed comparison of minimal requirements for promotion in the other Gulf universities.)

Slight differences exist in the minimal requirements for promotion in the Gulf states (El-Shibiny, 1997, pp. 168–169; see Appendix). However, in all Gulf states both a doctoral degree and a publications record are generally required.

Table 9.2   Promotion requirements in Saudi Arabian universities[a]

Rank	Teaching experience	Publication record	Evaluation criteria	
Associate professor	4 years since last appointment, at least 1 at a Saudi university	4 publications, 2 single-authored	Publications[b]	60 points
			Teaching	25 points
			Service[c]	15 points
			Total	100 points
			Decision by majority (2:1)	
Professor	4 years since last appointment, at least 1 at a Saudi university	6 publications, 3 single-authored	Publications[d]	60 points
			Teaching	25 points
			Service[c]	15 points
			Total	100 points
			Unanimous decision[e]	

Notes
[a] Council for Higher Education (1998). *Regulations*, arts. 21–22, 25–28, 32–33.
[b] Of which at least 35 points must be obtained.
[c] To the university and community.
[d] Of which at least 40 points must be obtained.
[e] If one of the three Promotion Committee members is dissenting, the evaluation of the publication record is passed to a fourth referee whose evaluation is final.

Increasingly, teaching performance is debated as a pertinent issue for promotion. In Saudi Arabia, for example, teaching performance accounts for 15 percent of the total points required for promotion to the rank of associate or full professor. A publication record accounts for the greater share in the promotion process.

Nonnationals may seek promotion based on the same procedure applicable for nationals. Yet, nonnationals may also apply for the recognition of a promotion obtained in a non-Saudi university (e.g., a home university), provided the home university is "recognized" by the host university.

Teaching

Teaching constitutes the major on-campus activity of Gulf faculty members. A "teaching unit" consists of a semester-long weekly "theoretical lecture" of 50 minutes, or a semester-long weekly "practical" or "field" lesson of 100 minutes. Faculty members with the rank of language teacher are expected to teach 18 weekly units (or 15 hours); instructors and lecturers are expected to teach 16 units (13 hours). Assistant professors are expected to teach 14 units (12 hours), associate professors 12 (10 hours), and full professors 10 (8 hours). Thus, the teaching loads of laboratory technicians, research assistants, instructors, language teachers, and lecturers are significantly higher than that of their counterparts holding the academic ranks of assistant, associate, and full professor. The latter three academic ranks are expected to devote "thirty-five weekly working hours—which could be raised to forty weekly working hours by decision of the University Council—to be spent on teaching, research, academic tutoring, library hours, [participation] in scientific committees and other works they are entrusted with by the competent university authorities" (Council for Higher Education, 1996, art. 41). Administrative functions—such as the duties of deans, deputy-deans, center directors, and department heads—entitle their bearers to a substantial reduction in teaching load.

The language of instruction is usually Arabic, in soft knowledge fields. However, in hard knowledge fields—such as the natural sciences, medicine, and science and technology—the medium of instruction is English (El-Shibny, 1997, pp. 163–164). In fields in which teaching requires the use of electronic equipment, such as Geographic Information Systems (GIS), English is used as well, because of the difficulty of finding references in Arabic, and in the region the software used is mostly in English (Al-Qaydi, 1999, p. 585). It should be observed that the Arabization of university instruction is not peculiar to Gulf universities but reflects a much-debated topic in higher education within the Arab states as a whole (Al-Ansari, 1988).

Teaching in Gulf universities encapsulates conflicting cross-cultural and cross-linguistic facets. For example, a study was undertaken of the teaching styles of nonnational faculty members and the learning styles of national students at Oman's Sultan Qaboos University, specifically during science lectures (Arden-Close, 1999). The study found that Western faculty members regarded memorization strategies, prevalent among Omani students, as being "deficient" and sought to introduce problem-solving strategies instead. Yet, the lack of cultural knowledge among Western lecturers, Arden-Close observed, further prevented them from devising pertinent and culturally adapted teaching strategies (see also Safi, 1986, pp. 429–430).

From this perspective, then, the indiscriminate introduction of evaluation tools used in Western universities in order to assess teaching and research practices in Gulf universities raises critical questions and "sensitive" issues (Shaw, 1996). And yet, within some Gulf universities the evaluation of faculty teaching by students has become an increasingly institutionalized practice (*Faculty Handbook*, 2000). In some universities, steering committees are instituted to review various facets of academic performance, teaching included (Safi, 1986; Sara, 1997). The evaluation of entire academic programs is undertaken, too, as part of accreditation, often by Western teams of experts (U.K. Team, 2000). In Saudi Arabia, university-based Academic Development Centers were recently established to promote teaching skills and evaluate teaching practices. The Academic Development Center at King Fahd University for Petroleum and Minerals, for example, organized a "Discussion Forum on Faculty Evaluation by Students" and a seminar on "Balancing Your Teaching and Research Efforts." The seminar announcement flyer stated:

> The academic environment is becoming much more competitive with the internationalization of education and the incorporation of new technologies in the learning process. In order to be an effective teacher, a faculty member needs to be up-to-date in his [*sic*] field. Based on his research, he can incorporate in the classroom many practical examples to share with students and bring the topic to life. He needs therefore to balance his efforts between teaching and research.

Various models for the evaluation of teaching have been explored in Gulf universities. For example, at Oman's Sultan Qaboos University, faculty members ranked administrators' evaluations of teaching as being the most effective compared to peer, self- and student evaluations (Albandary, 1996). Student evaluation of faculty teaching was perceived as being the least effective. Al-Imam Mohammad bin Saud Islamic University, in Saudi Arabia, introduced instructional programs to improve the quality of teaching (Al-Jameel, 1992). While both administrators and faculty members strongly agreed on the importance of the programs, they nonetheless perceived their contribution as being rather modest.

Furthermore, evaluating academic performance—especially teaching effectiveness—is not viewed as a neutral process and has, in fact, been described as a political activity (Shaw, 1996, p. 321). In Gulf universities, evaluation tends to be perceived as a "sensitive endeavor" and a "threatening activity" by officials and faculty members who feel that the privacy of their classrooms has been invaded (Sara, 1997, pp. 55–56; Al-Jameel, 1992). Moreover, evaluating academic performance may sometimes generate "pressures" with respect to outcomes. This is particularly so, as Gulf universities do not maintain a clear distinction between executive and legislative functions (Sara, 1997, p. 55).

Research

In Gulf universities a broad distinction exists between research in the fields of science and technology and research in the humanities and social sciences (including Islamic studies and *Shari'a*). Gulf universities and research centers were able to boost their research output significantly in the fields of science and technology. The share of Gulf universities and centers of research grew steadily during the 1967–1995 period, from 2.2 to 35.4 percent of all science and technology articles published by all Arab states combined (Zahlan, 1999).[2] Additional data reveal that the leading Gulf universities publishing in these fields were Saudi Arabia's King Saud University and King Fahd University for Petroleum and Minerals, with 18 and 14 percent, respectively, of all Gulf science and technology publications in 1995. Kuwait University followed with 9.2 percent. Thus, together, these three universities produced 41.2 percent of all Gulf-produced science and technology publications in 1995 (Zahlan, 1999, pp. 77–78, table 8.3). About one-quarter of all science and technology papers produced in Saudi Arabia and Kuwait were written with U.S. partners. This share was almost equal in size to all coauthored science and technology articles by Gulf and other faculty from all Arab states combined (Zahlan, 1999, p. 127).

According to Zahlan (1999, p. 69), the significant increase in science and technology productivity in GCC countries has been attributed to "wise national policies" that "attracted qualified research staff and provided them with a conducive work environment." As a result, many Arab, but also European and American, academics "left their national universities" and took positions at the expanding Gulf universities. Moreover, the Iraqi invasion and the dismantling of Kuwaiti research facilities in 1990–1991 did not prevent Kuwait University from substantially regaining its previous research output by 1995 (Zahlan, 1999, p. 69).

A study based on extensive interviews of Saudi Arabian faculty members in the fields of science and technology identifies four major factors accounting for what is termed "the poor performance of the Saudi universities in the area of scientific research" (Alghafis, 1992, p. 86). These are: lack of adequate research infrastructure, a stress on teaching and applied research, the limited "scope and practice" of research as a career, and the lack of "effective and useful linkages with the international scientific community and the network of international scientific institutions."

No comprehensive or comparative empirical studies were undertaken into social science and humanities research in Gulf universities. It has been reported that only a few of the research institutions operating in the Arab states as a whole directly support social science research. Rather, these institutions sponsor many seminars, conferences, and workshops involving local scientists. These forums encourage mainly the publication of policy-oriented papers and research communications. However, "the sense of balance within the discipline is lost when the bulk of the writing in it is produced for this purpose" (Shami, 1989, p. 651). The observation has been made that "it is probably fair to claim that generation of knowledge on social reality, and social change, is grossly deficient compared to what is required of science as a locomotive for human progress in the region"(Fergany, 2000).

Some partial and scattered primary data on Gulf social science and humanities research is available, subject to the limitations such data impose on any generalization. For instance, Kuwait University has been monitoring the periodic growth of faculty research output during the last decade. A *Research Abstracts* annual series is published for that purpose. Moreover, between July 1, 1997 and January 31, 1998, 142 research projects were awarded to faculty members at Kuwait University, of which 134 were sponsored entirely by the university (Kuwait University, 1998, pp. 6–8). Fifty-two percent of all projects were conducted in the fields of the social sciences and humanities. Of these, 35 percent were in the fields of Islamic *Shari'a, Hadith,* and *Qur'an* studies, a significant increase compared to previous years. Also at Kuwait University, the Implementation Office Committee, in December 2000 and January 2001, approved eight new research projects, two of which were in the social sciences and humanities. Increasingly, calls for multidisciplinary research are being made, "for studies that address prime social and community concerns." It was further observed that, so far, the "move towards an era of multidisciplinary and collaborative studies . . . [has] been rather limited due to an overwhelming concentration on pursuing studies of individual interests"(Ismaeel, 1998, p. 2).

A second illustration concerns Saudi Arabia. In 1998, the Council of Higher Education enacted a Unified List for University Research (ULUR) (Council for Higher Education, 1998). The ULUR prescribed the creation of a Deanship for Scientific Research in each university, seconded by a Deanship Council for Scientific Research. All research centers within the university were placed under the direct supervision of this deanship. The ULUR represents an attempt to coordinate and promote research and publication activities within Saudi universities. It further stipulated the establishment of the necessary mechanisms to expand research activities in the humanities, social sciences, and natural sciences and technology. At the present time, the ULUR is too recent a development to allow evaluation of its effects on university research and publications productivity.

University Libraries

University libraries serve as important teaching and research support facilities. Yet, in the case of Gulf universities, it appears that libraries are caught between contradictory processes. On the one hand, libraries in the region have short histories, having been founded between 1963 and 1986. Where gender separation is the norm, such as at the University of Qatar and UAE University, separate library services were established for men and women, with all the ensuing duplication in management, staffing, and titles acquisition. In other libraries separate areas were allocated to women to avoid duplication of collections, services, staffing, and budgetary expenses (Zehery, 1997, pp. 21–22, 28). Most Gulf university libraries operate as "administrative units... rather than as a primary educational component of the university," which negatively affects acquisition policies and collection development, as well as student and faculty usage of library facilities for both teaching and research. Furthermore, interlibrary loan programs and transfer of materials across the region and between institutions remain relatively limited (Zehery, 1997, pp. 21–22, 23, 28, 32–35). By and large, university libraries are still dependent on nonnational staffing. Locally trained professional librarians are scarce, in spite of the availability of several graduate programs in library studies (Ashoor and Chaudhry, 1999; Alghafis, 1992, p. 86).

Demographic Realities

Social formations, as well as state–university relations, pervade the academic workplace and reflect the intensity of social and political conflicts catapulted into Gulf universities.

Women have only limited access to power positions within the university system and spend much more time in administration compared to research. For instance, in Saudi Arabia, universities are gender segregated; their situation has been described as follows: "women faculty are professionally isolated, and female administrators are overburdened with teaching and administrative responsibilities while most lack pre- and in-service management training" (El-Sanabary, 1992, p. 13).

Some Gulf governments have taken actions to facilitate women's access to executive positions within the university system. For instance, in Kuwait, the ruling establishment has attempted—since the early 1990s, following the Gulf War—to expand the political and occupational opportunities available to women, in spite of persistent parliamentary opposition. In part, such policies were aimed at consolidating regime legitimacy, whether locally or internationally. In 1994, the Emir of Kuwait appointed a woman, a professor at Kuwait University and sister of Kuwait's National Assembly speaker, to preside over the country's only public university, a precedent for the Arab states (Bollag, 1994). A woman was later appointed deputy-minister of higher education, though with limited ministerial responsibilities. Similar official acts followed in Oman and Qatar as well, with the appointment of women to the position of deputy education minister.

In an attempt to explain these developments, with respect to Kuwait, it has been observed that

> the spread of education among members of formerly disadvantaged groups in Kuwaiti society promoted the development of a new class of able and ambitious Kuwaitis who did not come from the old powerful families.... [A]t first mostly male, [t]hey became managers in the state's oil company and professors at Kuwait University, both non-executive positions. [Yet,] to maintain its own position of dominance, ... [the powerful merchant class] was ready to change Kuwaiti social mores so that elite women of the merchant class could take over prestigious positions that might otherwise go to men from the emerging social classes. [As a result, w]orking women from the merchant class took professional positions in firms, government offices, and the university. (Tétreault and al-Mughni, 1995, pp. 409 and 410)

Thus, two major Kuwaiti social groups meet, as colleagues, within Kuwait University: women from the dominant classes and ruling elites and mobile men from less powerful classes and, increasingly, men from a Bedouin background as well. Clearly then, gender and class interact in the broader organization of the Gulf academic workplace. The extent to which such an interaction affects academic mobility and promotion is not clear at this point. No further data are available to probe the implications of this development, or to establish its existence across faculties and Gulf universities.

Differential terms of employment for tenured nationals versus contract nonnationals constitute one of the more conflict-laden dimensions affecting informal relations in the Gulf academic workplace (Al-Khalifa, 1990). Nonnationals are perceived as transient members within university systems eager to promote the representation of nationals. Nonnationals have a particularly high turnover rate. Yet, they have greater access to senior university, departmental, and professorial positions, particularly in the newer university systems in Bahrain, Qatar, Oman, and the UAE ('Abdallah, 1994, p. 132; Al-Farsi, 1997, pp. 187–188). Moreover, nonnationals play a significant role in the academic promotion of nationals, particularly in science and technology fields. Nonnational faculty members perceive themselves as the "real founders" and operators of Gulf universities ('Abdallah, 1994, p. 132). Yet, their national colleagues often complain that nonnationals have "their own priorities rather than the national interest at heart" (Al-Saadi, 1997, p. 194), whether with respect to staff development (Al-Farsi, 1997, pp. 187–188) or regarding the promotion of nationals. The latter "could mean replacement of expatriates, leading to a feeling of threat, the creation of conflict and a weakening of the trust between the local and expatriate" (Al-Saadi, 1997, p. 194). Within this context, nonnationals often feel marginalized and alienated ('Abdallah, 1994, p. 132), while nationals "consider this present policy to be frustrating their ambitious aspirations and therefore a strong cause for dissatisfaction" (Al-Saadi, 1997, p. 194).

Academic Freedom

Academic freedom in Gulf universities—as in the case of their Arab counterparts—remains a critical issue, affecting performance in the workplace. With respect to Arab academe in general, the Arab academic has already been likened to "a matador without a muleta" (Sabour, 1988, pp. 206–217). While being invested with the proper insignia to rank and status, Arab academics, as well as their Gulf colleagues, are often left with little effective power (Shaw, 1996). The patronage-client relations and a patriarchal social order have been described as powerfully pervading the Arab and Gulf academic workplace. These actively affect the relative career opportunities available to faculty members of different gender and class backgrounds (Sabour, 1988 and 1996; Shaw, 1997). Moreover, in many countries no explicit legal provisions are provided to secure academic freedom.

More particularly, organizational centralization and hierarchical relations constitute a major impediment to academic freedom, ultimately operating as a control mechanism. Often, the legislative and executive functions within Gulf public universities are not clearly delineated. Nor is there a clear separation of power between them, making it "particularly difficult to avoid undue pressure" (Sara, 1997, p. 57). The general observation has been made that "bureaucratic constraints in the financial management of grants and allocations restrict both the scope and breadth of scientific and technological research in the Gulf States' universities" (Alghafis, 1992, p. 28). UAE University has been described as suffering from "internal administrative oppression which hinders academic work" ('Abdallah,

1994, p. 129). The same source further stated that employment conditions of contract faculty—constituting the majority of the professoriate—operate as an additional constraint on academic freedom.

Other sources suggest that self-censorship and the avoidance of what are perceived as "sensitive issues" are widespread practices. Teaching, particularly in the humanities and social sciences, is not less devoid of value-laden controversies and ambiguities—whether political (Assiri, 1987), sociocultural (Moosa, 1999), or pedagogical (Obeidat, 1997; Haggan, 1999). For example, politically motivated actions were taken by university authorities against professors at Bahrain University (Committee on Academic Freedom, 1997). In Kuwait, on two recent occasions, cases were brought against university professors by ideologically minded groups, on grounds of defaming religion (Bencomo, 2000). These cases represent examples of a wide array of formal and informal social and political constraints affecting the work and academic performance of Gulf faculty members.

Gulf academics—as is the case with their Arab counterparts—are caught between restrictive or coopting government politics and the difficulty of asserting a critical and alternative discourse, with respect to political activism or nationalist or religious ideologies (Al-'Awwami, 1999). This double bind in which academics are caught—between state cooptation and limitations imposed by social mores and traditions—has become more pronounced in the aftermath of the Gulf War, particularly in Saudi Arabia (Nevo, 1998; Okruhlik and Conge, 1997).

A poem composed by 'Abdallah Hamid Al-Hamid expresses the issues at stake. Al-Hamid, a Saudi Arabian assistant professor from Riyadh, was arrested for political reasons in 1993. Released, he had to commit himself "to withdraw from all political activity considered hostile to the kingdom" (Gresh, 1996). His poem, "They Have Forbidden the Word," expresses the academic's deep despair and frustration when expression of free thought and practice are barred: [3]

> They have forbidden the word, writing and speech!
> Be silent! And if injustice remains
> When the tongue is mute, it will burn like a moth in the flame.
> For opinion now is trash, secreted away and thrown in the bin.
> The word is crime,
> Beware, he who would start a debate.

The issue of academic freedom is not limited in its implications to Gulf public universities alone, in terms of inhibiting social and leadership roles of faculty members. Rather, to varying degrees, it may also affect the feasibility of private higher education joint ventures between Western and Gulf organizations. For instance, the University of Virginia reportedly abandoned plans to open a branch in Doha, Qatar, at the invitation of the Qatari government, due to apprehensions regarding "human rights for women and members of religious minorities" (Smith, 2000). More recently, plans to open an extension of a Cornell University–sponsored Medical College in Qatar, fully funded by the Qatar Foundation (see the earlier section on "Gulf Private Higher Education"), were debated in Cornell's faculty senate. Though the plan has already been approved, the ensuing debate is nonetheless worth noting. A supporting senator asserted that this initiative "allows the medical school to have an international presence in that region of the world" making it "an extension of higher education on an international basis, part of a general movement in higher education in that direction." However, another senator expressed her concerns "about academic freedom and the human rights and safety of Cornell students and faculty." She further added that "[W]e cannot provide or assume that there will be the same kind of academic experience and rights as Cornell students here have with these restrictions" (Powers, 2001).

In his reply, Cornell's university counsel stressed that under Cornell's agreement with the Qatar Foundation, "the university has full operational autonomy," and that "from the earliest discussions with Qatar, Cornell considered the issues of academic freedom, personal safety and non-discrimination to be critical." He further revealed that the university has "consulted with the U.S. State Department, and in light of increasing globalization they were supportive of this initiative in the interest of educational diplomacy." Finally, he observed that recently "Qatar has considerably relaxed human rights restrictions" and that democratic elections were introduced, "although the government is still a conservative regime" (Powers, 2001).

CHARACTERISTICS OF THE PROFESSORIATE

The 13 Gulf public universities surveyed in this chapter (see table 9.1) employed over 11,000 teaching faculty members in 1998 and enrolled over 220,000 students.

Table 9.3 presents a breakdown of major faculty characteristics by gender, citizenship, academic rank, credentials, graduating period, and disciplinary orientation for each GCC member state.

Saudi Arabia has the largest university system in the Gulf, with about three-quarters of all Gulf faculty. Women constitute slightly over 15 percent of all Gulf faculty. Though significant differences

Table 9.3   Characteristics of Gulf university faculty members by state (percentages)[a]

Characteristics	Bahrain	Kuwait	Oman	Qatar	S. Arabia	UAE	All Gulf
Women	19.8	16.0	15.4	25.5	14.8	9.3	15.2
Citizenship[b]							
National		53.6	30.1	51.9	64.1	22.2	57.3
Arab		22.2	29.8	44.4	21.3	59.2	24.0
Western		11.6	25.2	2.8	2.6	14.5	4.9
Other		12.6	14.9	0.9	12.0	3.8	13.9
Rank[c]							
Nonacademic	40.6	46.7	42.8	55.1	30.6	52.4	34.9
Junior academic	37.1	31.8	29.1	25.2	37.2	25.6	35.3
Senior academic	22.3	21.5	28.1	19.7	32.2	22.0	29.8
Highest degree							
Bachelor's, Diploma	17.5	—	2.0	8.1	18.4	—	14.8
Master's	24.5	—	24.1	8.3	13.7	2.2	12.5
Doctorate	58.0	100.0	73.9	83.6	67.8	97.8	72.7
University attended							
Gulf	4.6	0.2	7.3	7.9	33.3	29.9	26.1
Arab	14.4	12.8	8.6	37.0	17.6	66.8	17.8
Western	72.4	77.5	45.5	51.9	45.9	1.8	50.8
Other	8.5	9.5	38.6	3.3	3.2	1.6	5.2
Graduation period							
Up to 1972	15.1	7.3	6.9	7.9	4.6	4.8	5.5
1973–1982	31.7	31.9	24.6	15.1	21.8	20.4	22.8
1983–1990	50.8	31.7	28.1	35.5	40.9	36.0	39.5
1991–1998	2.4	29.0	40.4	41.5	32.6	38.7	32.3
Disciplinary fields							
Arabic lang. & lit.	9.0	3.2	2.7	4.4	7.6	3.2	6.7
Islam & Shari'a[d]	—	6.1	1.2	9.5	16.2	4.3	13.3
Humanities	8.8	7.6	17.8	12.0	10.4	7.5	10.3
Education	11.1	9.1	9.8	10.2	5.5	11.2	6.7
Social sciences	15.7	20.7	12.0	22.0	15.5	24.0	16.7
S&T	49.2	39.6	45.7	41.0	33.2	38.8	35.4
Medicine & health	6.2	13.6	10.8	0.9	11.6	11.0	11.1
Total percent	100.0	100.0	100.0	100.0	100.0	100.0	100.0
Total faculty (*N*)	388	929	409	432	8,322	626	11,106
Relative % in Gulf	3.5	8.4	3.7	3.9	74.9	5.6	100.0

Notes
[a] The data on which table 9.3 is based pertain to the year 1998. They were derived from several sources: university faculty directories, official publications and data of the Association of Arab

Table 9.3 Continued

Universities (1999), and Internet sites of the various universities and regional organizations. Based on these sources, an integrated database comprising individual-level data of 11,106 faculty members was built. The database refers to over 90 percent of all Gulf *university* faculty. The table refers to the 13 public universities listed in table 9.1, with the exception of King Khalid University, for which data were not available.

[b] Though no data on the citizenship status of faculty members in Bahrain were available, it bears noting that the absolute majority of faculty members in this state are nonnationals. For practical ends, the term "Western" countries encompasses Europe, North America, Australia, and New Zealand.

[c] *Nonacademic* ranks included lecturers, assistant lecturers, teachers, assistant teachers, language teachers and research assistants. *Junior academic* ranks included assistant professors. *Senior academic* ranks included all associate and full professors. This categorization permitted the comparison of ranks across states.

[d] The University of Bahrain did not list any Islamic studies and *Shari'a* faculties or departments.

exist among states, Saudi Arabia and the UAE have lower percentages of women. The reason is primarily related to the recruitment of women into teaching positions within nonuniversity tertiary settings, such as gender-specific colleges. In Saudi Arabia, universities and colleges taken together, women constitute about 35 percent of all Saudi Arabian faculty (Mosa, 2000). In the smaller Gulf states, particularly Qatar and Bahrain, the relatively higher representation of faculty women is due primarily to the presence of nonnational faculty members.

Gulf universities differ in their reliance on nonnational faculty. About 57 percent of all Gulf faculty members are citizens of the Gulf state in which they work, the remaining part being citizens of other states. Interestingly, only a negligible number of those faculty members who are citizens of any particular Gulf state work in another Gulf state (less than 0.08 percent of the total).

The proportion of nonnational faculty members varies considerably among states. The lowest proportion is found in the older universities, such as Saudi Arabia (about one-third). The proportion is higher in the newer universities, as in the case of the UAE (almost 80 percent) and Bahrain (about 90 percent). In Kuwait, the percentage of nonnational faculty members declined, within 20 years, from about 83 percent in 1979 to about 47 percent in 1998 (Al-Ebraheem and Stevens, 1980, p. 216). About half of Qatar's faculty is still nonnational. Slightly over two-thirds of faculty members in Oman are nonnationals, compared to 90 percent in the mid-1990s (El-Shibiny, 1997, p. 168).

Less than a quarter of all Gulf faculty members are citizens of another non-Gulf Arab state, predominantly Egypt. The presence of Arab (non-Gulf) citizens is much more felt in the universities of Bahrain (the majority), the UAE (59.1 percent of the total in this country), Qatar (over 44 percent), and Oman (almost 30 percent).

Saudi Arabia and Kuwait have slightly over one-fifth each. These figures are indicative of the role played by Arab and, more particularly, Egyptian faculty in the expansion of Gulf universities. For many Egyptian academics, Gulf universities offer attractive occupational opportunities, particularly given the more limited economic and mobility opportunities within their home universities. The migration of Egyptian faculty has also served as a significant carrier of ideological movements from within Egypt into Gulf societies, student movements, and universities (Al-Siddani, 1985).[4]

Almost a fifth of all Gulf faculty members are citizens of states outside the Arab region as a whole, representing Asian (e.g., India, Pakistan), Western (mainly the United States and the United Kingdom) and several sub-Saharan African countries (e.g., Ghana, Nigeria, South Africa). Yet, citizens of Western countries account for less than 5 percent of the total number of faculty members; in the majority of cases, they are naturalized Arabs or individuals of Arab background. Citizens of Western countries are more predominant in Oman (25 percent), the UAE (almost 15 percent), and Kuwait (almost 12 percent).

For states more reliant on a nonnational academic workforce, it is possible to observe higher proportions of teaching appointments made to the lower academic ranks. This may be due to a higher representation of recently recruited nationals who start their tenure-track career at the lower positions. The lower academic ranks also include nonnationals employed as laboratory instructors or English-language teachers. In the case of Saudi Arabia, the professoriate is more evenly distributed across ranks.

Doctorates are held by almost three-quarters of all Gulf university faculty. Kuwait and the UAE have the highest percentage of doctorate-holding faculty members, while Bahrain has the lowest percentage (58 percent). Notwithstanding, over half of all Gulf faculty members are graduates of Western universities, in the absolute majority of cases universities in the United States and United Kingdom. Only about a quarter of all Gulf faculty members have graduated from Gulf universities. In Kuwait and Bahrain, around three-quarters of the faculty are graduates of Western universities and about an additional 14 percent have graduated from other Arab universities (e.g., Egyptian). With respect to Qatar, over one-third of its faculty graduated from Arab universities, and about 52 percent from Western universities.

Clearly, then, the Gulf professoriate is professionally trained and socialized in two major, and quite different, university environments— namely in Western (mostly the United States and the United Kingdom) and Arab (mostly Egypt) countries. In this sense, different academic and scholarly traditions ultimately find their way into the Gulf academic workplace and coexist, often within the same faculties and departments. How such an encounter affects departmental cooperation and collegial relations, let alone promotion opportunities, is not clear at this time. Further research is needed into these institutional aspects of the Gulf academic workplace. Nevertheless, these figures suggest that Gulf universities still rely heavily on Western, and to a lesser extent on Arab, universities as well for the training of their future academic cadres, particularly in hard knowledge fields.

The Gulf professoriate is relatively young. Over one-third of all Gulf faculty members graduated in the aftermath of the 1991 Gulf War, during the 1991–1998 period. An additional 39.5 percent graduated during the 1983–1990 period, under conditions of economic recession and declining oil prices following the "oil boom" decade. Yet, compared to other Gulf states, Qatar, Oman, and the UAE—and to a considerable extent Saudi Arabia, as well—have a relatively young academic workforce, with around 40 percent of their professoriate having graduated during the post–Gulf War period alone. These three states have further preserved a systematic growth in terms of the percentage of younger faculty joining the university. By contrast, Bahrain had the lowest growth rate in new faculty members, with only 2.4 percent graduating in the aftermath of the Gulf War.

Gulf states also differ in the distribution of their faculty across disciplinary fields. While the emerging hard knowledge fields now encompass almost half of all faculty members, meaningful differences can still be observed among states. For instance, a comparison between Oman and Saudi Arabia is pertinent in this respect. Oman has the largest concentration of faculty members in hard knowledge fields, 56.5 percent of all its faculty. However, Oman also has the largest percentage of faculty members involved in the humanities and education combined, 27.6 percent. At the same time, Oman has the lowest percentage, less than 4 percent, of faculty members teaching Arabic language and literature, and in particular Islamic studies (including Islamic jurisprudence, or *Shari'a*). By contrast, about a quarter of all Saudi Arabian faculty teach in the fields of Arabic and Islam, while less than 16 percent teach in all other humanities and education-related fields combined. Moreover, Saudi Arabia has a much lower percentage of its faculty engaged in science and technology. These contrasts between Oman and Saudi Arabia suggest that the Gulf academic workplace differs across countries in terms of the curricular and intellectual experiences provided for students and faculty alike.

It bears observing that an additional examination of the data (not presented in table 9.3 above) revealed that Gulf citizens tend to be found more particularly in soft knowledge fields, and much less in hard knowledge fields. In the latter, Gulf universities rely, some more dramatically than others, on a nonnational academic workforce. Where Arab citizens and graduates from Arab and Gulf universities are found, they tend to be concentrated in Islamic studies and *Shari'a,* the humanities, social sciences, and Arabic language and literature. Where Western and other nationals and graduates are found, they tend to be concentrated in science and technology and the medical sciences. This pattern clearly indicates that the "nationalization" policies of the academic workforce have remained relatively more effective in soft knowledge fields, particularly Islam-related fields of studies.

CONCLUSION

The present chapter examines the organization of the academic workplace and the characteristics of the professoriate in Gulf public universities. Gulf universities emerged and continue to expand in the crucible of comprehensive social, political, and economic transformations. Oscillating among local patronage relations, regional political rivalries, state-building policies, and globalized economies, Gulf universities have faced conflicting pressures that have also affected the organization of the academic workplace and the characteristics of the professoriate.

Increasingly drawing on the American academic experience, particularly in the areas of curricular structure and contents, teaching practices, and research and development policies, Gulf universities remain nonetheless entrenched in and organized around a patriarchal social order. While exhibiting some organizational and managerial features of their Western counterparts, Gulf public universities differ significantly from the former (and to no lesser extent among themselves) in terms of their practical institutional arrangements and organizational cultures. In spite of the increasing adoption of entrepreneurial, competitive, and research-oriented discourses, the Gulf academic workplace remains nonetheless steered by centralized agencies and top-down modes of governance. In many respects, the bureaucratic apparatuses governing the Gulf academic workplace are said to represent mechanisms of control and cooptation of emerging elites ('Abdallah, 1994; Bahgat, 1998 and 1999, p. 130; Sabour, 1988; Shaw, 1996 and 1997).

Tenure and collegial management constitute basic components of academic freedom. With respect to Gulf public universities, while non-nationals are on renewable contracts, both nonnational and national faculty members have only limited participation in the management of the university. In a centralized system of governance and administration, faculty participation largely supplements official initiatives and directives. Under such conditions, faculty participation in the decision-making process exhibits complex contradictions. Calls for faculty contribution to research productivity are increasingly being heard, particularly in the area of basic and multidisciplinary research. Similarly, the quality of teaching is increasingly perceived as a component of accountability and effectiveness of the university. While these policy trends imply greater autonomy and choice for both faculty and students, university governance and administration continue to operate primarily on the basis of state managerialist perspectives and priorities.

Relatively more attention has been granted to the political sociology of Gulf universities. Much less was done to probe how their internal human landscapes converge to shape the academic workplace. The present findings indicate that the Gulf academic workplace, as a context and space of practice, is characterized by significant sociodemographic and institutional cleavages that ultimately determine "how work is divided and done, how it is scheduled, supervised, compensated, and regarded by others" (Johnson, 1990, p. 1).

Ostensibly, the gender construction of the Gulf academic space is the most immediately discernible characteristic. As has been discussed earlier, within Arab academe, women are a subordinate minority in an overwhelmingly male faculty body, although significant differences exist between Gulf states (Sabour, 1988 and 1996). This is particularly visible if one compares, for instance, the gender-segregated Saudi Arabian universities and their coeducational Omani or Kuwaiti counterparts.

Similar gender-based divisions of labor have been referred to as a "patriarchal gender contract," ultimately reproducing broader gender roles (Moghadam, 1993). In Kuwait, Oman, and Qatar, where women have been appointed to senior university and ministerial positions, these policies rather express what has been called "Arab state feminism" (Hatem, 1995). In the same vein, male national faculty members are emerging as a class of civil servants. This emerging academic elite is dependent on state revenues and bound by broader power relations in its mobility prospects, whether within the university or beyond. Nonnationals constitute the overwhelming majority in the private sector and a significant percentage of faculty members in the universities. Thus, national female faculty members have to compete with both national men and nonnational labor in their bid for academic employment and promotion within the academic workplace. The distinction between nationals and nonnationals constitutes a significant line of demarcation among faculty members. It further fuels institutionally based identity politics and differential role expectations. For distinct groups of nonnationals, Gulf universities have become an appealing venue of long- or short-term migration (depending upon the group concerned), in view of economic and professional mobility. The greater concentration of Arab citizens and graduates of Arab (including Gulf) universities in Islamic and Arab studies, humanities, and the social sciences; and the greater concentration of citizens of Western countries and graduates of Western universities in science and technology and the medical sciences affect the relative share of these different citizenship-based groups within the reward system and their degree of involvement in organizational planning and decision making.

In Qatar, Oman, and the UAE—as well as in Saudi Arabia—younger graduates are increasingly being enrolled, as faculty members, in soft knowledge fields and relatively less in hard knowledge fields. Younger faculty members are unequally distributed, not only across ranks but also across disciplines. In a patriarchal and patronage-based society, the ability of the younger cohorts of faculty members to formulate and promote a teaching and research agenda based on an alternative vision to that of the established sociopolitical order or of dominant social groups is likely to involve a complex exercise in power relations. Younger generations of faculty are entering a career-oriented academic path, in increasingly unstable economic and political conditions, particularly in the aftermath of the Gulf War. Others have already observed that the expansion of Gulf higher education is associated with rising political dissent and opposition. They suggested that in Saudi Arabia, the expansion of state-sponsored Islamic higher education institutions has been associated with the rise of political contestation of the established political order, with greater demands being exerted on state-entrenched elites (Dekmejian, 1994; Nevo, 1998; Okruhlik and Conge, 1997). Similar processes were identified in other Gulf states. In Kuwait, the increasing entry of young academics from the "new social groups" composed of Islamists, *Shi'a* Muslims, and Bedouins has been associated with the rise of opposition politics (Tétreault and al-Mughni, 1995).

Clearly, then, the Gulf academic workplace is challenged simultaneously by both local and regional sociopolitical constraints and increasingly globalized markets and international migratory flows. Yet, the ultimate outcomes of these processes are far from converging toward a uniform organizational pattern of the academic workplace. Rather, an array of diverging institutional formations may be clearly identified. For instance, in Saudi Arabia, the expansion of Islamic studies and jurisprudence (*Shari'a*) was accompanied by a more circumscribed yet quite significant parallel growth in science and technology fields. Such a structural "bifurcation," it may be argued, reinscribes within the academic workplace the symbolic markers from which ruling elites within the Saudi Arabian state secure and reproduce regime legitimacy, albeit not without resistance. In contradistinction, the academic workplace in both Oman and Bahrain is constructed around what appears to be a more secular(ized) version of academe, immersed in Arab culture in Bahrain and science and technology, humanitites, and the arts in the case of Oman.

The Gulf academic workplace may thus be conceived as a "site" producing a myriad of hybrid organizational and institutional arrangements. From that particular perspective, the Gulf academic workplace is not a mere uniform outcome of global, regional, and local realities. It rather mediates what has been called "new constellations and contradictory spaces" (Helvacioglu, 2000, p. 340). As such, it reflects differential versions of situated modernities, elaborated in relation to both context-specific factors and regional global processes.

Such a conclusion requires the development of a research approach going beyond the simplistic dual conception of the Gulf academic workplace as being associated with transition from tradition to modernity or merely related to state-building and human-capital formation. Rather, it becomes imperative to probe how global discourses about academe are ultimately mitigated by symbolic local practices and power politics, which then surreptitiously (re)define the borders and institutional arrangements characterizing the academic workplace.

APPENDIX

Minimal requirements for academic promotion in Gulf universities

University	Requested rank	Degree	Time constraints	Academic productivity	Remarks
UAE UAE University	professor	doctorate	5 years as assistant professor or 10 years since doctorate	publications, teaching, or socially distinguished activities	special provisions may justify reduction of the time period by up to 2 years
	assistant professor	doctorate	5 years as lecturer	publications, teaching, or socially distinguished activities	
	lecturer	doctorate			
Ajman University for S&T	professor	doctorate	10 years following doctorate	3 publications during tenure as associate professor	3 years of experience as researcher, teacher, or expert
	associate professor	doctorate	5 years following doctorate	3 publications	3 years of experience as researcher, teacher, or expert
	assistant professor	doctorate			3 years of experience
	teacher	first or second degree			

Appendix (*continued*)

University	Requested rank	Degree	Time constraints	Academic productivity	Remarks
Universiy of Sharjah	professor	doctorate	5 years at previous rank	8 research-based publications at previous rank	
	associate professor	doctorate	5 years at previous rank	5 research-based publications at previous rank	
	assistant professor	doctorate			
	lecturer	master's			
Bahrain University of Bahrain	professor	doctorate	5 or 10 years	at least 36 points for research or 15 points for individual publications	time since last appointment may be reduced by one year
	associate professor	doctorate	5 years as assistant professor	at least 22 points for research (highest in the institution) and at least 5 points for single-authored publications	at least 10 years of experience in field, following second degree, 6 years as lecturer
	lecturer I	master's	10 years of expertise in field, after second degree, 6 years as lecturer	At least 16 points for published research	

University	Rank	Degree	Experience	Publications	Other
Arabian Gulf University	professor	doctorate	4 years at previous rank	5 single-authored publications or 9 coauthored published or accepted works	evaluation is point-based
	associate professor	doctorate	4 years at previous rank	5 single-authored publications or 3 coauthored published or accepted works	
	assistant professor	doctorate			
Oman Sultan Qaboos University	professor	doctorate	3–4 years at previous rank	expertise in field and published works	
	associate professor	doctorate	3–4 years at previous rank	expertise in field and published works	
	assistant professor	doctorate	2–3 years at previous rank		
	teacher	doctorate or master's			5 years teaching and research experience at a university
Qatar University of Qatar	professor	doctorate	6 years at previous rank or 12 years since doctorate, 1 spent at UQ	8 original research units, 6 published in refereed journals	

Appendix (*continued*)

University	Requested rank	Degree	Time constraints	Academic productivity	Remarks
	assistant professor	doctorate	6 years since doctorate, 1 spent as university teacher	6 original research publications, 3 in refereed journals	
	lecturer	doctorate			
Kuwait Kuwait University	professor	doctorate	4 years at previous rank, 14 years since doctorate	10 publications in scientific journals; 5 in the humanities and social sciences (5 single-authored)	candidate may be appointed without a doctorate on condition of having obtained a first degree 20 years earlier and a record of relevant publications
	associate professor	doctorate	4 years in previous rank		candidate should attend professional workshops in field of expertise and

				teaching workshops
assistant professor	doctorate	9 years since first degree, 5 years as lecturer	5 published works, 3 single-authored	candidate may be appointed without a doctorate on condition of having obtained a first degree at least 15 years earlier and having relevant publications
lecturer	doctorate	4 years since doctorate		candidate may be appointed without a doctorate on condition of having obtained a first degree at least 8 years earlier and having relevant publications

Source: Association of Arab Universities, on-line database accessible at www.aaru.edu.jo. For Saudi Arabia, see general information displayed in table 9.2.

NOTES

1. Information on salaries was derived from the following sources: for Kuwait, *General information,* 1996; for the United Arab Emirates University, *Faculty Handbook;* for Saudi Arabia, Council for Higher Education, 1996 and n.d., refer to *Regulations* and *NS-Regulations* and particularly the salary scales in these two documents; for Oman, Al-Saadi, 1997, p. 201; for a general description and discussion, Al-'Areed, 1993.

2. Zahlan (1999) surveyed published science and technology articles indexed by the Philadelphia-based Institute for Scientific Information.

3. The poem was originally composed in Arabic. It was first translated into French and appeared in the August 1995 issue of *Le Monde Diplomatique,* in an article on Saudi Arabia written by Alain Gresh. The article was subsequently translated into English and appeared in *Index on Censorship* (Gresh, 1996), from which the present version is reproduced.

4. For illustration of the radical changes that took place at the level of the professoriate, the case of King Saud University (KSU), Saudi Arabia, is of interest. In 1970, there were 163 faculty members employed at KSU (then, Riyadh University): 45 Iraqis, 24 Egyptians, 22 Saudi Arabians, 19 Pakistanis, 18 Jordanians, 13 Syrians, 7 Palestinians, 7 British, 4 Indians, 2 Americans, 1 Algerian, and 1 Australian (Abd-el-Wassie, 1970, p. 57). According to the present data, in 1998, there were 2,224 teaching faculty members at KSU (or one-fifth of all faculty members in Saudi Arabia). Saudi Arabian nationals constituted 13.5 percent of all faculty members in 1970, compared to 55 percent in 1998. Arab nationals constituted 66 percent of all KSU faculty in 1970, compared to 35 percent in 1998. Egyptians constituted 22 percent of all Arab nationals at KSU, compared to 56 percent in 1998.

REFERENCES

'Abdallah, 'Abd El-Khaleq. (1994). Al-hurriyat al-acadimiya fi Jami'at Al-Imarat Al-'Arabiya Al-Muttahida [Academic freedoms in the United Arab Emirates University]. *Al-Mustaqbal Al-'Arabi* [The Arab future], 190, 121–134.

Abd-el-Wassie, Abd-el Wahab. (1970). *Education in Saudi Arabia.* London: MacMillan.

Al-Ansari, Muhammad Jaber. (1988). Al-ta'rib al-jami'i wa-hatmiyat al-muqaraba al-maydaniya [University Arabization and the necessity of field orientation]. *Risalat Al-Khaleej Al-'Arabi* [The Arab Gulf Newsletter], 24, 151–189.

Al-'Areed, Jaleel Ibraheem. (1993). *'Adhu hay'at al-tadris bi-jami'at duwal al-Khaleej al-'Arabiya: Ta'hileh wa-taqwimeh* [The teaching faculty member in the universities of the Gulf states: His training and evaluation]. Riyadh, Saudi Arabia: Arab Bureau of Education for the Gulf States.

Al-Assad, Nasser Al-Din. (1996). *Tasawwurat Islamiya fi al-ta'lim al-jami'i wal-bahth al-'ilmi* [Islamic conceptions of university education and scientific research]. 'Amman, Jordan: Rawae' Majdalawi.

Al-'Awwami, Faisal. (1999). *Al-muthaqqaf wa-qadaya al-din wal-mujatma'* [The intellectual and issues of religion and society]. Beirut, Lebanon: Muntada Al-Kalima Lil-Dirasat Wal-Abhath.

Albandari, Mohammed S. (1996). The perceptions of faculty and administrators in the Teacher Education Colleges and the Sultan Qaboos University in Oman about faculty evaluation. Unpublished doctoral dissertation, New Mexico State University, Las Cruces.

Al-Ebraheem, Hassan Ali, and Stevens, Richard P. (1980). Organization, management and academic problems in the Arab university: The Kuwait University experience. *Higher Education,* 9(2), 203–218.

Al-Farsi, Fawziya. (1997). Omanization and faculty development in Sultan Qaboos University. In K. E. Shaw (Ed.), *Higher education in the Gulf: Problems and prospects* (pp. 182–191). Exeter, U.K.: Exeter University Press.

Alghafis, Ali Nasser. (1992). *Universities in Saudi Arabia: Their role in science, technology and development.* Lanham, MD: University Press of America.

Al-Jameel, Saleh Adbulrahman. (1992). Administrators and teaching staff perceptions toward instructional development program at Imam Mohammad bin Saud Islamic University, Saudi Arabia. Unpublished doctoral dissertation, University of Pittsburgh.

Al-Karni, Ali Saad M. (1995). Evaluating the performance of academic department chairpersons. *Higher Education,* 29(1), 37–57.

Al-Khalifa, Fatma Hasan. (1990). *The impact of organizational structure on university administration: The case of Kuwait University.* Unpublished doctoral dissertation, George Washington University.

Al-Misnad, Sheikha. (1985) *The development of modern education in the Gulf.* London: Ithaca Press.

Al-Qaydi, Saif Salim. (1999). Teaching GIS in the Gulf Co-operation Council universities. *GeoJournal,* 47, 583–586.

Al Rawaf, H. S., and Simmons, C. (1992). Distance higher education for women in Saudi Arabia: Present and proposed. *Distance Education: An International Journal,* 13(1), 65–80.

Al-Saadat, Abdullah, and Afifi, Elhami. (1990). English via closed-circuit television in a sex-segregated community. *British Journal of Educational Technology,* 21(3), 175–182.

Al-Saadi, Khalifa. (1997). Job satisfaction amongst Omani staff in Sultan Qaboos University. In K. E. Shaw (Ed.), *Higher education in the Gulf: Problems and prospects* (pp. 192–203). Exeter, U.K.: Exeter University Press.

Al-Siddani, Nouriya. (1985). *Al-jama'at al-daghita: Al-kuwa al-tulabiya al-Kuwaitiya* [Pressure groups: the Kuwaiti student movement]. Kuwait: Dar Al-Siyasa.

Al-Sulayti, Hamad. (2000). Education and training in GCC countries: Some issues and concerns. In *Education and the Arab world: Challenges of the next millennium* (pp. 271–278). Abu Dhabi, UAE: The Emirates Center for Strategic Studies and Research.

Arden-Close, Christopher. (1999). Conflict of learning styles: University science lectures in the Sultanate of Oman. *Journal of Science Education and Technology*, 8(4), 323–332.

Ashoor, Mohammed-Saleh Jamil, and Abdus Sattar Chaudhry. (1999). *The education of library and information professionals in the Arabian Gulf region.* London and New York: Mansell.

Assiri, Abdul-Reda. (1987). Teaching political science in the Gulf states: The case of Kuwait. *International Studies Notes*, 13(3), 78–81.

Bahgat, Gawdat. (1998). The silent revolution: Education and instability in the Gulf mornarchies. *Fletcher Forum of World Affairs*, 22(1), 103-112.

———. (1999). Education in the Gulf monarchies: Retrospect and prospect. *International Review of Education*, 45(2), 127–136.

Bencomo, Clarisa. (2000). *Promises betrayed: Denial of rights of Bidun, women, and freedom of expression.* New York: Human Rights Watch.

Berkey, Jonathan. (1992). *The transmission of knowledge in medieval Cairo: A social history of Islamic education.* Princeton, NJ: Princeton University Press.

Bollag, Burton. (1994, February 16). A female president, the Arab world's first, guides the restoration of Kuwait U. *Chronicle of Higher Education*, A45.

Buraimi varsity to grant degrees. (2001, November 18). *Oman Daily Observer.* Retrieved from www.omanobserver.com.

Caesar, Judith. (1999). A view from the Gulf. *The Progressive*, 63(2), 24–25.

Committee on Academic Freedom in the Middle East and North Africa. (1997). Dismissal of Dr. Al-Zeera and Lack of Academic Freedom in Bahrain. *Middle East Studies Association Newsletter*, 19(2).

Council for Higher Education. (1996). *See* Saudi Arabia. Ministry for Higher Education. Council for Higher Education. (1996).

———. (1998). *See* Saudi Arabia. Ministry for Higher Education. Council for Higher Education. (1998).

———. (n.d.) *See* Saudi Arabia. Ministry for Higher Education. Council for Higher Education. (n.d.).

Dekmejian, Richard Hrair. (1994). The rise of political Islamism in Saudi Arabia. *Middle East Journal*, 48(8), 627–643.

Del Castillo, Daniel. (2001, September 7). First private university is set to open in Oman. *Chronicle of Higher Education.*

El-Sanabary, Nagat. (1992). *The Saudi Arabian model of female education and the reproduction of gender divisions.* Working Paper No 16. Los Angeles: University of California, G. E. Grunebaum Center for Near Eastern Studies.

El-Shibiny, Mohamed. (1997). Higher education in Oman: Its development and prospects. In K. E. Shaw (Ed.), *Higher education in the Gulf: Problems and prospects* (pp. 150–181). Exeter, U.K.: Exeter University Press.

Faculty handbook (2000). Al-'Ayn: United Arab Emirates University, Office of the Deputy Vice Chancellor for Academic Affairs.

Fergany, Nader. (2000). Science and research for development in the Arab region. In Eglal Rached and Dina Craissati (Eds.), *Research for development in the Middle East and North Africa*, Chap. 2. Ottawa: International Development Research Centre. Retrieved from www.idrc.ca/books/focus/930/13fergan.html.

General information for faculty members (in Arabic). (1996, September). Kuwait: Kuwait University, Office of the Vice President for Academic Affairs.

Gilbar, Gad. (1997). *The Middle East oil decade and beyond: Essays in political economy*. London: Frank Cass.

Gresh, Alain. (1996). End of an era. *Index on Censorship*, 25(4), 46–54.

Haggan, Madeline. (1999). A linguist's view: The English department re-visited. *Forum*, 37(1). Retrieved from *www.exchanges.state.gov/forum/forum.htm*.

Halloran, William F. (2000). Zayed University: A new model for higher education. In *Education and the Arab world: Challenges of the next millennium* (pp. 323-330). Abu Dhabi, UAE: The Emirates Center for Strategic Studies and Research.

Hatem, Mirvat. (1995). Political liberalization, gender, and the state. In Rex Brynen, Bahgat Korany and Paul Noble (Eds.), *Political liberalization and democratization in the Arab world* (pp. 187–208). Boulder, CO: Lynne Rienner.

Helvacioglu, Banu. (2000). Globalization in the neighborhood: From the nation-state to Bilkent center. *International Sociology*, 15(2), 326–342.

Ismaeel, Assad A. (1998). RA envisions an era of multidisciplinary research. *The Researcher* [Bulletin of the Research Authority at Kuwait University], 6(1), 1–2.

Johnson, Susan Moore. (1990). *Teachers at work: Achieving success in our schools*. New York: Basic Books.

Kuwait Islamists victorious in segregation battle. (2000, June 13). *Gulf-News.com*. Retrieved from www.gulf-news.com.

Mahdi, Kamil. (1997). Some economic aspects of higher education in the Arab Gulf. In K. E. Shaw (Ed.), *Higher education in the Gulf: Problems and prospects* (pp. 19–41). Exeter, U.K.: Exeter University Press.

Makdisi, George. (1981). *The rise of colleges: Institutions of learning in Islam and the West*. Edinburgh: Edinburgh University Press.

Moghadam, Valentine. (1993). *Modernizing women: Gender and social change in the Middle East*. Boulder, CO: Lynne Rienner.

Mograby, Abdullah. (2000). Human development in the United Arab Emirates: Indicators and challenges. In *Education and the Arab world: Challenges of the next millennium* (pp. 279–307). Abu Dhabi, UAE: The Emirates Center for Strategic Studies and Research.

Moosa, Samira M. (1999). New trends in home economics in the Sultanate of Oman. *Journal of Family and Consumer Sciences*, 91(5), 11–14.

Mosa, Ali A.(2000). Pressures in Saudi Arabia. *International Higher Education*, no. 20, 23-25.

Nevo, Joseph. (1998). Religion and national identity in Saudi Arabia. *Middle Eastern Studies* 34(3), 34–53.

Obeidat, Marwan M. (1997). Language vs. literature in English departments in the Arab world. *Forum*, 35(1). Retrieved from www.exchanges.state.gov/forum/forum.html.

Okruhlik, Gwenn, and Conge, Patrick. (1997). National autonomy, labor migration and political crisis: Yemen and Saudi Arabia. *The Middle East Journal*, 51(4), 554–565.

Powers, Jacquie. (2001). Faculty Senate discusses professorial titles and medical college in Qatar. *Cornell Chronicle, 32 (No. 31)*, 19 April. Retrieved from www.news.cornell.edu//Chronicles/4.19.01/FacultySenate.html.

Qatar Foundation commits $750 million to establish Weill Cornell Medical College. (2001). *Fund Raising Management* 32(4), 13.

Rida, Muhammad Jawad. (1998). Al-thaqafa al-thalitha: Al-jami'at al-'Arabiya wa-tahadi al-'ubur min barjakh al-thaqafiyin [The third culture: Arab universities and the challenge], *Al-Mustaqbal Al-'Arabi* [The Arab future], 237, 109–120.

Sabour, M'hammed. (1988). *Homo academicus Arabicus*. Joensuu, Finland: University of Joensuu.

———. (1996). Moroccan academic women: Respectability and power. *Mediterranean Journal of Educational Studies*, 1, 75–92.

Safi, A. Qayum. (1986). Kuwait University and its evaluation program. *Higher Education*, 15(5), 421–447.

Sara, Nathir. (1997). Internal evaluation in higher education: Toward a model for Third World countries. In K. E. Shaw (Ed.), *Higher education in the Gulf: Problems and prospects* (pp. 45–68). Exeter, U.K.: Exeter University Press.

Saudi Arabia. Ministry for Higher Education. Council for Higher Education. (1996). Regulations governing the affairs related to Saudi university employees from among faculty members and those under their authority (in Arabic). Decision 4/6/1417. Riyadh: Ministry of Education.

———. (1998). Unified list for scientific research in the universities (in Arabic). Decision 2/10/1419. Riyadh: Ministry of Education.

———. (n.d.). Regulations for the employment of non-Saudis in the universities (in Arabic). Riyadh: Ministry of Education.

Shami, Seteney. (1989). Socio-cultural anthropology in Arab universities. *Current Anthropology*, 30(5), 649–654.

Shaw, K. E. (1996). Cultural issues in evaluation studies of Middle Eastern higher education. *Assessment and Evaluation in Higher Education*, 21(4), 313–324.

———. (1997). Reforms in higher education: Culture and control in the Middle East. In K. Watson and S. Mogdil (Eds.), *Educational dilemmas: Debate and diversity*, vol. 2 (pp. 206–218). London: Cassell.

Smith, Thomas W. (2000). Teaching politics abroad: The internationalization of a profession? *Political Science and Politics,* 33(1), 65–73.

Tansel, Aysit and Kazemi, Abbas. (2000). Educational expenditure in the Middle East and North Africa. *Middle Eastern Studies,* 36(4), 75–98.

Tétreault, Mary Ann, and al-Mughni, Haya. (1995). Modernization and its discontents: State and gender in Kuwait. *Middle East Journal,* 49(3), 403–417.

U.K. team assesses men's college chemical engineering programme. (2001, March 4). *Gulf-News.com.* Retrieved from www.gulf-news.com.

Zabalawi, Isam H. (2000, 25–26 November). *Academic freedom and gender equities.* Paper presented at the International Conference on Higher Education in Asian Universities: Challenges and Future Trends. Sharjah, UAE: University of Sharjah.

Zahlan, Antoine. (1999). *Al-'Arab wa-tahadiyat al-'ilm wal-tiqana: Taqadom min dun taghyir* [The Arabs and the challenges of science and technology: Progress without change]. Beirut, Lebanon: Markaz Dirasat Al-Wihda Al-'Arabiya [Centre for Arab Unity Studies].

Zehery, Mohamed H. (1997). University library development in the Arab Gulf region: A survey and analysis of six state university libraries. *International Information and Library Review,* 29, 13–44.

1 0

THE ACADEMIC WORKPLACE IN A CHANGING ENVIRONMENT: THE NIGERIAN SCENE

Monica Iyegumwena Barrow and
Fidelma Ekwutozia Ukeje

Academic staff in Nigerian universities perform the traditional basic functions of teaching, research, and public service. However, the country's unique history and rapid political and economic changes have affected the academic structure, institutional governance, and resources and finances of Nigerian universities, which have tended to follow different trajectories than have those in Europe, America, and elsewhere. Years of military rule under various regimes have had an impact on institutional leadership and the federal government's commitment to higher education, resulting in inconsistent higher education policies. Currently, the government is promoting the economic philosophy of deregulation, liberalization, and privatization and as a result cannot adequately fund education. The underfunding has placed educational systems at a crossroads: Various unions in higher education institutions go on strike for a greater part of the year, and the infrastructure is inadequate and deteriorating. To complicate matters, the number of universities and student enrollments continue to increase, which has resulted in the inefficient duplication of program offerings in the universities. Poor salaries and deteriorating conditions of service have led to a mass exodus or "brain drain" of academic staff and helped to bring about poor leadership in the institutions. All these factors have had a serious impact on the academic profession and the environment for teaching and research; the result has been low academic productivity on the part of both staff and students.

The Context

Over the last three decades, the federal government of Nigeria, recognizing the important role of university education in nation building, has increased the number of universities and expanded their programs in order to meet the challenges posed by the growing economy. Through its laws, it has attempted to ensure uniformity of standards and quality at all institutions of higher education. It has consistently provided the universities with about 95 percent of their operating costs. However, in the mid-1980s, with the downturn in the global economy and the resultant decline of revenue from oil, which is the mainstay of the nation's economy, it became clear that the government could no longer cope with the increasing demand for university education.

With the weakening of the government's position as the major funding source, the gap between the demands of universities for funding and the grants provided gradually widened. Yet the universities were not allowed to charge tuition fees. This decline in funding adversely affected the quality of teaching and research, as well as the general working conditions in the universities. Universities in Nigeria are compelled to operate under adverse conditions: poor salaries and conditions of service; lack of resources for nonsalary academic expenditures, such as textbooks, journals, teaching, and research equipment, and maintenance of such equipment; overcrowding and deteriorating physical facilities; and erratic electricity and water supplies.

The deteriorating working conditions and poor salaries have led to the mass exodus of academic staff from the universities to other countries and into the private sector of the Nigerian economy, where conditions of service are more attractive. The field that suffers most from brain drain is the medical sciences. In addition, few individuals coming out of postgraduate schools wish to lecture in the universities, because of the poor working conditions. This has left many institutions with young, inexperienced, and inadequately trained staff who lack the necessary mentors and role models to guide them (Saint, 1992). Links between Nigerian and overseas universities for the purpose of teaching, research, staff development, and student exchange programs have been greatly curtailed. Most academics are unable to remain current in their fields because of a lack of equipment and facilities, especially computers and access to information and communications technology. Attendance at local and international conferences, seminars, and workshops is almost impossible because of lack of funds.

The report of the Presidential Commission on Salary and Conditions of Service of University Staff (1981) and that of the Commission on Review of Higher Education in Nigeria (*Higher Education in the Nineties,* 1991), as well as a recent background study conducted on the design of the Nigeria University System Innovation Project by the World Bank (Dabalen, Oni, and Abekola, 2001), attest to the deteriorating conditions inside the universities. The situation has resulted in a lowering of academic standards and of the quality of graduates. Graduates are deficient in written communication and technical proficiency, which makes them unfit for the labor market (Dabalen, Oni, and Adekola, 2000).

The lack of good leadership and outmoded, inflexible managerial structures are also problems confronting the universities. Quite a number of university staff are given high-level administrative jobs as vice-chancellors, deans, heads of academic departments, and research administrators with little or no prior administrative training or experience. Appointed mainly for their academic qualifications, administrators have been generally criticized for their lack of managerial skills, which has affected the management of universities and by implication the management of academic matters. Financial constraints have also affected research activities in the universities; for quite some time, academics have complained of inadequate funding for research.[1] The challenge confronting universities, therefore, is to improve the deteriorating conditions of the academic workplace in order to enhance teaching and research activities.

While most of the Nigerian university system is comprised of federal, state, and private universities, this chapter will focus primarily on the federal universities. However, most of the issues discussed are common to all universities, irrespective of ownership.[2]

EXPANSION AND GROWTH OF UNIVERSITIES

University education in Nigeria dates back to 1948, when University College, Ibadan was established by what was then the British colonial government. The university was affiliated with the University of London and became a full-fledged university in 1962. Following the report of the Commission on Post-School Certificate and Higher Education in 1960, known as the "Ashby Commission," four additional universities were established. By the mid 1970s, the number of universities had increased to 13, to meet the higher education demands of the emerging nation. With the goal of increasing the level of participation of its citizens in education, the government has continued to create new educational institutions. By 2000, the number of federal universities had risen to 25 (5 of which were universities of technology, 3 universities of agriculture, and 1 federal military university), up from 6 universities in 1970. At present, in addition to the federal universities, there are 14 state-owned universities, and more are in the pipeline. The 1993 National Minimum Standards and Establishment of Institution Decree for the first time allowed private-sector participation in the provision of higher education. Presently, there are three privately owned universities. The government is considering applications for the establishment of more private universities.

The increase in the number of universities has resulted in a proliferation of academic disciplines in the university system. A 1985 comparative study on certain Nigerian universities and some of the newer British universities revealed the relatively high level of diversification of academic programs in Nigerian universities (Oduleye, 1985). The study showed that while nearly all Nigerian universities tended to stretch their resources over all academic groupings, their British counterparts did not. Some academic programs are considerably underutilized throughout the Nigerian university system.

Another phenomenon in Nigeria is the high demand for access to universities. This is because university education is perceived as the most important instrument for the development of a specialized workforce needed for nation building. It is also regarded as a mechanism for upward mobility in Nigerian society, which accounts for the desirability of a university education. The demand for access has remained high and exceeds what the higher education system can accommodate. This has deprived many who are qualified of access to a university education. For example, in 1998, only 35,000 students were admitted out of the 350,000 who applied. Although the Nigerian university system is the largest in Africa, with a student population of about 400,000, it has been argued that Nigeria's participation rate in higher education is one of the lowest in the world (Jibril, 2000). The participation rate was 667 per 100,000 inhabitants in 1995, compared with South Africa's 1,524, North America's 5,544, and Europe's 3,285. By 2010 it is predicted that there will be 22 million Nigerians between 18 and 25 years of age. If only 50 percent are eligible for higher education, and the university system continues to absorb 56 percent of those pursuing higher education, then Nigerian higher education institutions on the whole would need to admit about 11 million students, with the universities accommodating 6.2 million of the new entrants. This points to the need to encourage the development of more private

Table 10.1 Student enrollment by gender

Year	Male (%)	Female (%)	Total
1965	6,922 (89.8)	787 (10.2)	7,709
1970	12,394 (85.7)	2,074 (14.3)	14,468
1975	27,148 (84.1)	5,138 (15.9)	32,286
1980	60,692 (78.0)	17,099 (22.0)	77,791
1985	103,243 (76.0)	32,540 (24.0)	135,783
1990	132,016 (73.0)	48,855 (27.0)	180,871
1995	240,652 (71.7)	95,138 (28.3)	335,790
1998	257,048 (64.3)	142,764 (35.7)	399,812

Note: From *Higher education in the nineties and beyond,* 1991; and National Universities Commission, forthcoming.

universities and the provision of distance-learning facilities. Currently, out of about 400,000 students in all Nigerian universities, the federal universities alone account for a total number of 267,730 students.

As table 10.1 shows, the total student population rose to 399,812 in 1998 from 7,709 in 1965. This increase in students has greatly impacted the nature of the academic profession in the areas of infrastructural provision, equipment, and academic staff. Learning and teaching conditions have deteriorated, as universities were built without adequate consideration given to the facilities required to run academic programs. Subsequently, to control growth and guarantee quality of output and cost-effective use of resources, universities were grouped according to their institutional age and level of development. There are currently three recognized university "generations," with differing annual growth rates that form part of the funding parameters used in allocating resources from the government to the universities: first generation, 2.5 percent; second generation, 10 percent; and third generation, 15 percent.

The gender distribution in enrollment shows that the proportion of female students in the universities has been quite low, although there has been a steady increase over the years. In 1998, women constituted only 35.7 percent of the total student population in the federal universities. The disparity is attributed to a number of factors—such as social, economic, and cultural prejudices against women in the society—that put women at a disadvantage (*Higher Education in the Nineties,* 1991). These low female enrollments have had an effect on academic staffing ranks, in particular, and human resource needs of the nation, in general.

GOVERNANCE AND CONTROL

In Nigeria, education, especially at the tertiary level, is a federal government monopoly that has witnessed an evolution in government intervention (Federal Ministry of Education, 1989). The government has subjected the university system to intense scrutiny through committees, visitation panels, and commissions. Through these organs, recommendations have been made on how best to strengthen and run the universities. At the national level, the Federal Ministry of Education is charged with the primary responsibility of formulating educational policies for the Nigerian university system.

As part of the process of effecting federal control over the universities, two organizations were established, the National Universities Commission (NUC) and the Joint Admissions and Matriculation Board (JAMB). The NUC was set up in 1962 following the recommendations of the Ashby Commission, with the primary objectives of ensuring orderly development of university education in Nigeria, maintaining standards, and ensuring adequate funding. The NUC was patterned after the British University Grants Commission, now the University Funding Commission, to act as a buffer between the universities and the government. The Minimum Standards Decrees of 1985, 1989, and 1993 vested in the NUC the power to lay down minimum academic standards for all the universities in the federation (irrespective of ownership) and to enforce these standards. The power to accredit university degrees and other academic awards was also conferred on the NUC. These statutory provisions were made to ensure the quality of education and the uniformity of education practice in all universities in Nigeria.

Academics have criticized the NUC for its inability to secure adequate funds for the universities and for causing delays in disbursing secured funds. They have also resisted the setting of minimum academic standards. Yet the NUC has been empowered to lay down minimum academic standards.

The other coordinating body for the universities is the JAMB, established in 1977 to conduct entrance examinations and to oversee placement of potential students at the various universities. However, this centralized form of control has attracted the criticism of academics, who regard the establishment of this body as an encroachment on their academic functions and in conflict with the norms and functions of institutional autonomy and academic freedom.

At the institutional level, the federal government has, through statutory provisions, ensured uniformity in administrative structures and practices in all the universities. There are assigned specified roles, clear hierarchies, and chains of command, which are essentially features of a bureaucratic model. The model, critics assert, is outmoded and inflexible and may inhibit the capacity of the universities to react creatively and swiftly to environmental pressures and opportunities (Davies, 1985). As mentioned earlier, outmoded and inflexible managerial structures have been one of the problems confronting the universities in Nigeria. The inability of universities to recognize and respond swiftly to the changing environment has led to increased control and intervention by the government to bring about desired reforms.

The university's administrative structure consists of a governing council appointed by the country's president (the "visitor") and headed by a pro-chancellor who is responsible for the general direction and control of the universities and answerable to the visitor through the minister of education. The 16 members of the council are drawn from outside and within the university and serve for three-year terms. The vice-chancellor is the chief administrative and academic officer of the university and is responsible to the governing council. The vice-chancellor is an academic, and in most cases a professor, appointed for a five-year term by the president from a list of three candidates nominated by the senate and confirmed by the council. In some cases, nominations from the universities are ignored, and vice-chancellors are appointed from outside the institution and imposed on the institution. The involvement of the government in the appointment of university administrators is viewed by some as an infringement on the institution's right to appoint council members. The appointment of vice-chancellors, in particular, has been a bone of contention between the university unions and the federal government. The principal officers—the registrar, bursar, and university librarian—are appointed by the governing council and are responsible to the vice-chancellor for the day-to-day administration of the university.

Within the academic structure, the senate is the highest academic authority. Membership includes the vice-chancellor (who serves as chair), deans of several faculties, directors of several academic institutes, professors, heads of academic departments, and the university librarian. The university senate oversees academic matters—including the organization and control of teaching, promotion of research, and the awarding of degrees, certificates, and diplomas. The academic board, which reports to the senate, is responsible for the planning, coordination, development, and supervision of academic work at the university. The board is composed of the vice-chancellor (who acts as chair), deans, and department and unit heads. Deanships of faculties are rotated among professors for two-year terms, and are determined by an election of peers in the faculty and departments. The department is the basic academic unit in the university, and is headed by a senior academic. Headships are rotated among senior academics in departments for two-year terms.

These institutional structures, however, have come under some criticism. Academics have often argued that at various times during the last two decades the military and civilian heads of state as "visitors" have exercised powers to discipline students or staff, to close or open the universities, or to appoint vice chancellors or other officials in violation of the provision of the law.

FINANCIAL PRESSURES AND ACCOUNTABILITY

The federal universities in Nigeria rely heavily on the government for grants, generating very little revenue internally from rental property, endowments, commercial ventures, and consultancy services. The government provides the universities with funds through a long-established formula based on full-time equivalents. Of this recurrent grant to the universities, 10 percent is set aside for library development and 5 percent for research. Salaries represent 60 percent of the recurrent grant. Forty percent of the main capital grant goes to teaching and research equipment.

From 1960 to the mid-1970s, Nigerian universities enjoyed healthy funding, mainly because they were few in number. In the early 1970s, during the oil boom years, the nation had no difficulty in sustaining university education through adequate funding. With the creation of more universities, however, government resources became overstretched. Coupled with this funding shortfall was the lack of an agreed upon and consistent policy for funding the universities. Funding appeared to be simply what the nation could afford. For example, based on NUC sources, in 1998 the universities' request to the government for recurrent grants amounted to N12,114,192,331 (U.S.$96,913,539),[3] but the funds released to universities amounted to just N4,523,747,549 (U.S.$36,189,980). The main capital grant for the same year was N807,910,767 (U.S.$6,463,286). However, within the same year, supplementary grants of N1,542,592,994 (U.S.$12,340,743) for recurrent expenses, N187,951,500 (U.S.$1,503.612) for capital (refurbishment) expenses, and N573,855,666 (U.S.$4,590,845) special capital expenses (refurbishment) were released to the universities. The total release to universities amounted to N8,502,905,980 (U.S.$68,023,247). In real terms, considering the high inflation and the low value of the naira, the allocations were quite low. This continued low-level funding has forced expenditures per student down from U.S.$700 in 1991 to U.S.$362 in 1998. The minimum internationally recommended unit cost is U.S.$1,000 per annum—whereas in North America, the unit cost per student is U.S.$5,596; in Europe, U.S.$6,585; and in South Africa, U.S.$1,058 (Jibril, 2000). The low funding has also led to misuse of scarce funds by some university administrators, with a negative impact on the academic workplace.

The NUC has instituted a process for monitoring the utilization of funds released to the universities. It ensures that funds for research, library development, and teaching and research equipment are not diverted to other ventures by insisting that the universities properly account for previous releases of funds before new releases are made. The release of research grants, for example, did not occur during the years they were allocated because some universities failed to account for grants released the previous years. For the 1993–1996 period, only 53 percent of research grants were released, with just 14 universities receiving their grants through 1995 (National Universities Commission, 1998). These funding delays have had implications on the quality and quantity of research activities by academics in the universities. Generally, many academics complain that research grants to universities are inadequate for sustaining any serious research work. It is, therefore, important for administrators to encourage their academic staff to seek research funds from sources other than the government. This is essential, as quality teaching and research are expected from these institutions.

QUALITY CONTROL

With the rapid expansion of universities and the increase in enrollments since the 1980s, government and institutional leaders have been preoccupied with the issue of quality standards and assurance. The monitoring, control, and accreditation of academic programs in the universities take place in three main ways: by means of coordinating agencies, external examiners, and professional bodies.

The NUC is empowered to lay down minimum academic standards for all universities and other institutions of higher learning in the federation and to accredit their degrees in formal consultation with university academics. Minimum requirements are established for each program in such areas as admissions, course content, student and staff workloads, floor space for lectures, laboratory and library facilities, and student/staff ratios. The hope is that a student in any particular field of study will attain a certain minimum level of competency in that discipline.

Program visits to the institutions are carried out by accreditation panels, which are usually composed of professional and academic peers in the respective subject. While the members of the panel are largely drawn from the universities, others are from related professional bodies and associations. Deficiencies detected during visits are made known to the proprietor who is expected to take appropriate corrective action. Nonaccredited programs are subjected to periodic visits, timed to enable the proprietor to implement the accreditation requirements.

Accreditation exercises on Nigerian universities were carried out in 1991, 1995, and 1999. The government has since released the results of the 1991 exercise; the results of the remaining two have not yet been released. The 1991 results showed that even though the course contents of programs in most disciplines met the required standards, the programs were deficient in staffing and facilities. Out of a total of 836 programs, only 178 (21.3 percent) met all the required standards and received full accreditation status, while 575 (68.8 percent) programs received interim accreditation status and 83 (9.9 percent) were denied accreditation.

Since that exercise, the government intensified its efforts to improve the funding of universities. Through the 1990–1995 Nigeria/World Bank University Sector Adjustment Credit Facility, books, journals, and equipment were supplied to the federal universities. Predoctoral students were also trained under the project in order to increase the number of academic staff in the universities. The Education Tax Fund, which was introduced in 1993 and sets aside a substantial percentage of the fund for the improvement of university education in Nigeria, remains a potential source of relief to the university system.

External examiners are used in the universities in the final year of the program to assess courses and projects, and to certify the overall performance of graduating students as well as the quality of facilities and teaching. The external examiners send their written reports to the institutions concerned. Remunerations for this type of service are very minimal for the academic staff invited to serve as external examiners. However, with the poor salaries they receive, academics still look forward to carrying out this type of service.

As in many higher education systems, professional bodies in Nigeria play a role in a number of academic disciplines, regulating the teaching and practice of certain professions. Relevant bodies include the Institute of Chartered Accountants of Nigeria, the Institute of

Medical Laboratory Technology, the Council of Registered Engineers in Nigeria, Nigerian Medical Association, Pharmacists Councils of Nigeria, Nigerian Veterinary Medical Association, and the Council for Legal Education, to mention a few.

One other area that needs to be addressed to ensure academic quality in the university is teaching skills. Nigerian academics have been criticized for their poor teaching skills; this problem also affects primary and secondary schools (Fafunwa, 1971). The reason for this is the assumption that anyone who completes a Ph.D. will automatically be able to impart knowledge to others, which is not always the case. Nigerian academics should be encouraged to undergo some training in pedagogy.

All in all, the requirements for accreditation affect all aspects of the academic profession—quantity and quality of academic staff, quality and suitability of academic facilities and equipment, suitability of academic programs, and general conditions of service of staff members. The accreditation process generally prompts improvement in the academic workplace and should attract extra funding from the government.

THE ACADEMIC PROFESSION

The academic profession in Nigeria has been characterized by strife for many years due to the constellation of problems facing the academic workplace, as highlighted in the preceding sections. In the 1997–1998 academic year, there were a total of 16,591 academic staff members of all ranks in both federal and state universities (see table 10.2). In terms of academic ranks, professors and associate professors or readers constitute 14.5 percent of total academic staff; senior lecturers and senior research fellows, 23.2 percent; lecturers I and II and research fellows, 36.9 percent; and assistant lecturers and junior research fellows, 17.4 percent. Others, such as tutors and instructors, were 8 percent of the total.

Again in table 10.2, as indicated in the 1999 NUC Annual Report, the total number of academic staff at federal universities in 1999 was 13,800, compared with the NUC's approved academic staff strength of 23,806. This indicates a shortfall of 10,008. The figures in table 10.3 also indicate that professors/readers constituted 15.8 percent of the faculty; senior lecturers, 24.1 percent; and lecturers I and below, 60 percent. This contrasts with the approved ratio of 20 percent for professors/readers, 35 percent for senior lecturers, and 45 percent for lecturers I and below. This analysis reveals that the pyramidal structure for the academic staff in the universities was heavy at the base, and indicative of the brain drain in the entire system occasioned by

Table 10.2   Academic staff at federal and state universities, by rank and gender, 1997–1998

Rank	Male (%)	Female (%)	Total
Professors, associate professors, and readers	2,271 (94.2)	141 (5.8)	2,412
Senior lecturers and senior research fellows	3,395 (88.2)	453 (11.8)	3,848
Lecturers I and II and research fellows	5,199 (84.9)	924 (15.1)	6,123
Assistant lecturers and junior research fellows	2,447 (84.7)	442 (15.3)	2,889
Others (tutors, instructors, etc.)	1,126 (85.4)	193 (14.6)	1,319
Total	14,438 (87)	2,153 (13.0)	16,591

Note: From National Universities Commission, forthcoming.

Table 10.3   Actual and approved academic staff numbers, by rank and gender, 1997–1998

Rank	Actual	Approved	Shortfall
Professors and readers	2,178	4,761	2,548
Senior lecturers	3,327	8,332	5,005
Lecturers I and below	8,295	10,713	2,418

Note: From National Universities Commission, forthcoming—for the "actual" figures only; "approved" figures were derived by applying the NUC funding parameters.

the deteriorating working environment. To worsen matters, fresh intakes are not coming into the academic profession because of the poor conditions of service, so the quality of staff on the ground in the universities is further reduced.

With regard to academic staff distribution by gender, out of a total of 16,591 staff recorded in the 1997–1998 academic year (see table 10.2), 14,438, or 87 percent, were males while 2,153, or 13 percent, were females. The number and percentage of women professors have remained low, with a higher proportion at the lower levels—composing 5.9 percent of the professors, 11.8 percent of the senior lecturers, and 15.1 percent of the lecturers. This makes it difficult for female academic staff to have a significant impact as role models for girls, especially in the universities. Even though equal opportunity provisions are entrenched in the laws governing the universities, subtle discriminatory practices against women academics that may hinder their progress cannot be ruled out. There is a need to carry out research into the plight of women academics in Nigerian universities and the participation of women in university education in general to improve the present low level of female participation in university education and consequently the staffing levels in the universities.

No data are available on the age structure of academic staff. With few exceptions, universities in Nigeria have not yet encountered the phenomenon of the aging professor. One reason is the relatively young age of the universities; the oldest was established in 1948, and most of them date from between 1975 and 1995. It will be assumed that only a few of the professorial rank are of an average age of 55, while a majority of other grades of staff are under age 50. As mentioned earlier, many older and more experienced academic staff have left the universities for greener pastures.

In Nigerian universities, academic staff mobility is low.[4] Some universities are overstaffed in some disciplines, especially science and technology-based disciplines, while some are understaffed, particularly the newer institutions. The low mobility has been attributed to a number of factors—notably the tendency for tribal groups to dominate staff positions in the universities (Yesufu, 1973). It is a common and unfortunate phenomenon in Nigerian universities that staff are attracted mainly to institutions in their own localities. Staff sometimes fear that they will suffer from discrimination and unfair treatment if they move to universities that are not within their tribal region. Incentives will be required to encourage mobility within the university system in the country.

Full-time academic staff members are required to teach a minimum load of 8 credit units per semester, including postgraduate teaching. For science-based disciplines, this means a minimum of six lecture hours and two to three hours of laboratory work per week. For arts-based disciplines, this entails a minimum of six lectures and two one-hour tutorials per week. Classroom learning is based on traditional classroom presence and student tutoring by the academic staff. For instance, every full-time student is required to register for a minimum of 15 credit units per semester and a maximum of 24 credit units, except for students on field experience or industrial attachment. A student is expected to spend anywhere from four to six years to graduate depending on the program of study and not more than 50 percent more than the stipulated number of years to complete the program. A uniform course credit system has been adopted in the universities.

Regardless of these regulations, some academics carry a greater teaching load than they can accommodate, which is one reason their unions have confronted the government over the issue of their conditions of service.

Staff Unions

Although unions are not listed under the university law as components of the decision-making structure, their influence on decision making cannot be ignored. While academics around the world generally organize themselves into associations, in Nigeria university staff unions represent a departure from this norm. Faculty are organized in trade unions concerned with negotiating conditions of service with the government rather than with their immediate employers—the university councils.

The three main trade unions recognized in the universities for negotiations with the federal government are the Academic Staff Union of Nigerian Universities (ASUU), the Senior Staff Association of Nigerian Universities, and the Non-Academic Staff Union of the Universities. Like all other trade unions, the major objective of these unions is to maximize opportunities and security for their members—including higher standards of living, financial protection, academic freedom, and funding matters.

Over the years, a series of direct negotiations between university trade unions and the federal government for upward review of salary and nonsalary conditions of service often led to strikes and subsequent closures of universities whenever union demands were not fully met. For example, in 1975, the government's harmonization of salaries at universities with those of the civil service led to demands for a separate and higher salary as well as fringe benefits for academic staff in the system. Through the frequently used instrument of strikes, which should be a weapon of last resort in labor relations, the ASUU reached an agreement with the government in 1992 that brought about significant improvements in the conditions of service.

The recent demand by the union for a separate salary scale for academic staff would create a disparity between salaries of academic and nonacademic staff at the universities (as is the case in most parts of the world), a situation to which the unions representing nonacademic staff have objected. The reasons advanced by academics for a differential salary scale are that scholarship is a distinct profession with a complex function of teaching, research, and public service, and should be rewarded accordingly. They argue that academics are in very high demand internationally and that there is a shortage in Nigeria because of poor remunerations. The nonacademic staff, on the other hand, argue that the university system is made up of separate entities that perform distinct functions to achieve the desired results and therefore are interdependent. The government entered into negotiations with the ASUU on the issue of a separate and higher salary structure for academic staff. The nationwide ASUU strike lasted from April to July 2001. A threatened strike of nonacademic staff did not occur (Academic Staff Union, 2000). An agreement was reached between the ASUU and the federal government for a 22 percent increase in salaries of all academic and nonacademic staff.

The student unions, on the other hand, have often come out in support of their teachers' strike actions. On some occasions, the strikes were against repressive government policies, particularly during the military era. The universities have also witnessed student protests and strike actions against the deteriorating academic and student hostel facilities as well as the poor conditions of service of their teachers.

APPOINTMENTS AND PROMOTIONS

By the statutes that established them, Nigerian universities enjoy the freedom to appoint, promote, and discipline their staff. Academic positions are filled either by promotions from within the institution or by new appointments. Some universities advertise for positions when they do not have internal candidates to fill vacancies.

The basic academic qualifications for an academic staff member are a good first degree and a master's degree and a Ph.D. in the relevant discipline. (The bachelor's degree can be classified as 1st class, 2nd class lower, 3rd class or a pass.) An academic career starts at the lecturer II level. For promotion from lecturer II to lecturer I, the staff member should possess a Ph.D. degree and at least three years of experience on the job. In addition, the staff member will be assessed on the quality of teaching, publications, and contribution to the university and the community. To attain a senior lecturership, the staff member must have at least three years experience after attaining the lecturer I level, a significant number of publications, solid teaching experience, and participation in university administrative processes and community activities. To attain the rank of associate professor or reader, a senior lecturer must have at least three years of experience (as senior lecturer), outstanding achievements in research and teaching, a good number of publications resulting from research, as well as demonstrated academic leadership ability, coupled with community activities. External assessment of publications is also required. The position can be filled by promotion or appointment. To be promoted to the level of professor, the staff member must have had at least three years of university teaching experience as associate professor or reader, as well as a considerable number of publications, outstanding research work, postgraduate student supervision, teaching, and service to the university and to the community. External assessment of publications is required, as well as strong leadership ability. The reports by the external assessors on publications are submitted to the university's appointment and promotion committee.

Generally, this systems works but abuses do occur—for example, if the candidate happens to know the external assessors. However, a candidate must receive positive results from at least two of the external assessors in order to be successful. The service aspect of the candidate's qualifications varies across universities according to institutional mission and culture and includes service to a particular field in the form of consulting or committee work at the institutional and departmental levels.

All appointments for Nigerian academic staff are on a permanent and pensionable basis or what is often referred to as "tenure." The retirement age is 65 years for academic staff and 60 years for nonacademic staff. In the civil service the retirement age is 60 years. Previously, the retirement age was pegged at age 60 or 35 years of service, whichever came first, with a possible extension to 65 years on a contract basis.

"Tenure" implies that upon initial appointment the staff member must serve a required number of years, generally three years, on probation, becoming eligible thereafter for "confirmation" subject to a report of satisfactory performance in an annual evaluation by the department head and approval by the appointment and promotion committee. If after three years a person is refused confirmation, he or she might be asked to withdraw from the university. This practice provides job security and tends to reinforce academic freedom. Some critics, however, have argued that it lessens diligence, protects incompetent teachers, and limits opportunities for young teachers. Tenured status, once it is granted, can be terminated only on grounds of serious misconduct or incompetence.

There are observable shortcomings of the present tenure system as currently practiced in Nigerian universities. With poor salaries and unfavorable conditions of service, many lecturers engage in some other paid work outside the university, to enhance their income. This practice has the effect of relegating their academic responsibilities to second place. Many academics schedule their lectures for the evening hours to enable them to attend to their private businesses during the day. Since it is very difficult to weed out nonproductive staff because of their permanent appointments, except by natural attrition, many of them become noninnovative and make no tangible effort to update their knowledge in their fields. One very disturbing observation is that in some universities, research productivity is usually lower after the attainment of a professorial chair. Much of the research that is carried out is simply for the purpose of promotion. In the Western world, on the other hand, this is the stage at which most academics begin to fulfill their research potential and mentoring of younger scholars (Ipaye, 2000). Moreover, without an ongoing involvement in research, academics are often not kept abreast of current developments in their fields, which in turn affects the quality of the knowledge imparted to their students. This negative trend could, however, be reversed with the establishment of a reward system for research that made use of a market-orientation or links with industry and supported work generally relevant to national development needs. Tenure systems are currently undergoing modifications in some universities in other parts of the world. There is also the need in Nigeria to modify the tenure system at the universities in order to eradicate the identified deficiencies, stimulate scholarship, foster dedication to duty, and engender total commitment to the academic calling.

Nontenured appointments refer to those offered to retired academic staff who have passed the retirement age of 65 and are employed on contract. At some universities, honorary appointments are given to certain retired professors in recognition of their academic excellence during their active working lives. These academics do not receive emoluments but do enjoy certain privileges such as housing and use of academic facilities such as the libraries and laboratories. From time to time, they may be invited to deliver lectures. Appointments of non-Nigerians are also on a contract basis for a specific period and are renewable.

Few universities employ teachers on a part-time basis. The use of part-time staff is a new development at Nigerian universities and has not yet been fully explored. At present, part-time academic staff are mostly limited to part-time evening courses at universities located in urban centers. The inflexible career structures and conditions of service in the universities do not allow for part-time employment. The use of part-time academic staff could go a long way toward easing the present staff shortages in the universities.

For academic staff promotions, student evaluation of academic staff has become a component of staff assessment. Peer review, external reviews, research and publications, and teaching experience are other parameters.

The university's laws and statutes prevent termination of appointments before normal retirement age without "good cause." Typical causes for termination of appointments of a permanent academic staff member include voluntary resignation, mandatory retirement age, death, incapacity due to disability, neglect of obligations, and breach of canons or ethics prescribed for the particular discipline. Nevertheless, there have been instances when university staff including academic staff were removed in total disregard of established procedures for removal of staff. Such cases were common during the military era (CODESRIA, 1995).

Salary and Conditions of Service

As mentioned earlier, the salaries of academic staff were harmonized with those of the civil service in 1975, following the recommendations of the Public Service Review Commission. Prior to this, Nigerian universities enjoyed a system distinct from and somewhat more attractive than that which applied to the civil service and public corporations. Since harmonization, the university councils lost the power to negotiate the conditions of service with academic staff although the councils were in fact the employers. The salary and nonsalary conditions of service at Nigerian universities are now determined at the national level and uniformly applied without consideration of an individual university's circumstances or stage of development. The salaries are pegged to civil service pay scales, which are determined by the Wages and Salaries Commission. This situation creates a relatively inflexible system, which makes it difficult for universities to offer performance incentives or to reward excellence and does not encourage efforts to improve management efficiency (Saint, 1992).

The harmonization created a loss of morale and diminishing job satisfaction among the academic staff. The universities objected to their inclusion in the unified grading and salary structure and persisted in agitating for change. The government stepped in once again to resolve the issue by setting up the Presidential Commission on Salary and Conditions of Service of University Staff in 1981. This resulted in the introduction of the University Salary Scale. However, the salaries and conditions of service of staff in tertiary institutions worsened over the years relative to other sectors of the economy. Academic salaries were low and were declining in value owing to inflation and the exchange rate of the naira. The nonsalary conditions of service were likewise generally inadequate—with poor or nonexistent fringe benefits and depressing working conditions. Academics were perhaps the only social group not accorded civic recognition for their work, thus contributing to the diminishing status of academics in society. The overall situation has undermined the attractiveness of the academic profession, thus leading to an exodus into other sectors of the Nigerian economy and sometimes into jobs unrelated to the qualifications of academics.

Concerned about the exodus of senior academics, in 1992 the federal government set up a committee to review higher education in Nigeria. The committee brought some improvement to the situation of academics. In 1998, the government introduced a new national minimum wage, as well as the Enhanced University Salary Scale, which included improvements in the nonsalary conditions of service.

Due to an ongoing inflationary trend, however, the Nigerian government was not always able to meet its financial commitment to the higher education system, with the result that the general conditions at the universities and in the academic profession continued to deteriorate. Repeated crises in the universities and the strike actions by both staff and students sometimes resulted in lengthy closure of the universities (for example, the entire 1994 academic year). The situation became so serious that the government introduced the Harmonized Tertiary Institute Salary Structure (HATISS). Yet Nigerian academics still wanted their own salary scale as is the practice in other higher education systems.

However, as shown in table 10.4, under the HATISS staff salaries have recently increased substantially (by as much as 207 percent), along with increases in fringe benefits. In real terms, salaries are not competitive when compared with some other developing countries, such as South Africa and the West Indies (Academic Staff Union, 2000). The salary of the highest paid academic in Nigeria is N533,316 (U.S.$4,267) per annum—a situation that the Academic Staff Union claims is responsible for the mass exodus of Nigerian

Table 10.4 Annual basic salaries (HATISS)* of university academic staff

Rank	1999	May 2002
Graduate assistant	N64,596 (U.S.$517)	N198,816 (U.S.$1,591)
Assistant lecturer	N76,392 (U.S.$611)	N235,128 (U.S.$1,881)
Lecturer II	N87,984 (U.S.$704)	N270,780 (U.S.$2,166)
Lecturer I	N112,332 (U.S.$899)	N345,612 (U.S.$2,765)
Senior lecturer	N136,464 (U.S.$1,092)	N419,916 (U.S.$3,359)
Associate professor, reader	N154,632 (U.S.$1,237)	N475,788 (U.S.$3,806)
Professor	N173,328 (U.S.$1,387)	N533,316 (U.S.$4,267)

Note: From the Federal Government of Nigeria.
* Harmonized Tertiary Institutions Salary Structure.

academic staff and the low morale and diminishing job satisfaction of academics in the system.

Quite often, in making a comparative analysis, critics concentrate on the basic salary and fail to take into account allowances and other fringe benefits. In Nigeria, both academic and nonacademic staff receive many benefits, according to rank. These include a car, residential accommodation, housing allowance, leave allowance, free medical care, domestic servant allowance, transport allowance, entertainment allowance, and utility allowance, among others. Car loans, housing loans, and furniture loans at very reduced interest rates (3 to 4 percent) are also available to staff. Academic staff enjoy sabbatical leave of one year, with salary. They go on annual leave and are paid leave transport grants. A number of allowances apply only to academic staff: a journal allowance (12 percent of basic salary), a research allowance (6 percent of basic salary), an allowance for membership in a learned society (6 percent of basic salary), and an allowance for examination supervision (20 percent of basic salary). These allowances are meant to encourage academic staff to subscribe to scholarly journals in his or her area of specialization, carry out individual research, and participate in exam supervision. However, little serious research activity is carried out.

Some allowances—such as those for housing, transport, utility, entertainment, rent subsidy, and domestic servants—are paid along with the pension. There are also special provisions relating to the pension of professors. The Unions (Miscellaneous Provisions) Decree of 1993 provides that a person who has served a minimum of 15 years as a professor in the university and who "during the period of service was absent from the university only on approved national or university assignments" will receive a pension equal to his or her last annual salary, as well as any benefits to which he or she may be entitled. Staff members also earn a percentage of income generated through consultancy services.

To compensate for the dearth of teachers resulting from brain drain, especially in the sciences and engineering-based disciplines, the Nigerian Expatriate Supplementation Scheme (NESS) was introduced in 1988 for the federal universities. This allows the salary supplementation of expatriate academic staff recruited in the affected disciplines. Staff recruited from countries where the cost of living and salaries are lower than in Nigeria are not considered for supplementation. Supplementation is limited to the level of senior lecturer and above, the level of remuneration is tied to the nationality of the staff members. For example, a professor from the United States is paid an additional U.S.$13,000; from Europe, U.S.$9,000; and from Asia, U.S.$6,000. Payments are made in dollars and released to the staff members' overseas accounts. There are presently only 130 NESS beneficiaries, eight of whom are females. The scheme has not succeeded in attracting staff from North America and Europe, because of the great disparity between the earnings of academics in Nigeria and those in developed countries; it has mostly attracted expatriate staff from Asia and other African countries. Even with all the improved salaries and incentives, academics in Nigeria are still clamoring for higher salaries and a separate salary scheme. The solution to this lingering issue is for the government to empower the university councils to determine their individual university's conditions of service. In this way, various staff unions and individuals will negotiate directly with their councils on issues relating to wages, which will encourage competition and efficiency in the system.

POLICY ISSUES AND FUTURE TRENDS

Traditionally, Nigerian universities have enjoyed considerable freedom in the area of teaching and research. However, this cherished freedom has been threatened from time to time by the degree of control the government has had over university affairs, especially during the long military era, which created an adversarial relationship between the government and the universities. Changes in economic circumstances have also led to increased demand for accountability by the government.

The Academic Staff Union of Nigeria has continued to insist that academic freedom can be guaranteed only if universities are free from control in matters of their internal governance and finances (CODESRIA, 1996). Still, an ongoing issue is the extent to which universities can be allowed to administer themselves, particularly when they are funded by the government. Can autonomy work where universities are not self-sustaining?

As part of the demilitarization of society and full democratization of state institutions, the new democratic government is determined to remove all the bottlenecks that had hitherto denied the system its desired freedom. Therefore, considering the relevance of devolving its powers over the universities in line with the demands of academic freedom, the government is already working out a strategy for a smooth transition to a fully autonomous university system. A joint committee set up by the government (the Ijalaye Panel) is currently working on the modalities for putting the government's university autonomy policy into operation, including the preparation of draft legislation on the issue. It is expected that universities will have financial and administrative autonomy and will be free from all forms of administrative control from the government.

The governing councils will likely be the sole employers who can hire and fire. The councils will appoint or remove the vice-chancellor and determine the remuneration package and conditions of service of all university staff. Funds will be released as a block grant to each university, which each governing council will allocate according to the university's internal priorities. The senate of each university will determine the contents and details of curricula, and will be able to determine the criteria and modalities for the admission of students.

Besides these levels of freedom, autonomy means more and better funding for universities, which will strengthen them to be globally competitive. A fundamental policy position in this regard is the government's 40 percent increase of the recurrent allocation to universities in the 2001 budget. The intent was to help the government cope with the requirements of an enhanced salary and allowances scheme for all university staff and, of course, to prepare the ground for an autonomous university system.

Autonomy implies self-sustenance. Therefore the reintroduction of fees by universities is envisaged. This step may not cause student unrest because each university will be able to set its own fees, which will thus become a factor in competing with other institutions. Other changes will also be introduced. However, the government strongly believes that before fees can be reintroduced, Nigerians must be economically empowered to be able to pay such fees. That means that the economic situation must improve and the general living standard be raised to an acceptable level. Government has, in recognition of this, increased emoluments for all public servants and has introduced some far-reaching poverty alleviation schemes, particularly for rural citizens and other citizens who are not gainfully employed. It also reintroduced a federal scholarship scheme for undergraduates, beginning in 2001, to assist intelligent and indigent students. The sum of N1 billion (U.S.$8,000,000) was set aside for the project in the 2001 budget. States and local governments are also expected to take up the challenge and reintroduce bursary grants.

The government has introduced the Nigerian University System Innovation Program (NUSIP), which is intended to assist the universities in the following areas: improvement of teaching and learning by upgrading library holdings and instructional and research equipment; increasing institutional management capacities through management training; electronic networking; and special initiatives—including strategic planning, professional development (especially for academic staff), predoctoral training, and a graduate fellowship scheme for female students.

The professoriate under the NUSIP would benefit from improved infrastructural development and the provision of educational equipment, books, journals, and staff training to enhance their teaching and research capacities, thus enhancing their workplace environment. The federal government is now negotiating with the World Bank over support for the NUSIP in the form of a credit facility.

To increase access and remedy the shortfall in academic staff through long distance education, Nigeria plans to join the African Virtual University program in South Africa. The National Universities Commission is expected to coordinate the program in Nigeria using its Nigerian University Network as the platform for delivery. The government will provide the financial support for Nigerian universities' participation.

The government is also trying to develop a policy that will assist in the implementation of the Commonwealth of Learning program in Nigeria. These reforms and programs are seen as efforts to improve academic quality in the universities and restore the academic profession to its past glory. There are moves afoot by government to keep universities current on trends in higher education.

In the context of autonomy, a number of institutional changes are likely to occur in the future, including the devolution of authority to the university councils; an increase in revenue-generating activities; deregulation of conditions of service to allow flexibility, increased staff earnings, and improved working conditions; performance-related pay for academics; productivity-oriented incentives; and flexible career structures to allow part-time employment and mobility within the system.

With autonomy, the responsibility will fall on councils to determine remuneration packages and conditions of service for all categories of staff. It is expected that the universities will be able to retain their high-quality staff and attract others.

Academics want autonomy, but as has been stated, this has implications for academic freedom. Nigeria has traditionally enjoyed a considerable degree of academic freedom. However, during the period of military rule, universities were often not allowed to operate in accordance with their enabling laws, statutes, rules and regulations, with due process and within the laws of the land.

However, the granting of new autonomy is not without limitations. For example, to ensure transparency and accountability, the government will, for a while, continue to appoint the pro-chancellors and external members of the council. The management of autonomy will continually be monitored, with a view to further enhancement based on creditable performance. For quality assurance, accreditation of university programs will continue. The president will continue to be the visitor to the federal universities with the power to determine how often visitations are carried out—at least once every five years and more often if necessary.

At the time of this study, the bill on autonomy was awaiting passage by the National Assembly. Autonomy is viewed as a welcome development—a challenge to both universities and their leaders as the nation looks forward to greater improvement in institutional performance and government–university relations. Since the government will continue to be a major funding source, an improved government–university relationship is plausible. Government could participate in the definition of a university mission statement; such a strategic planning exercise would provide a framework for discussing the key questions about the university's future and could lead to a greater mutual understanding between the government and universities and hence agreement on autonomy and accountability (Saint, 1992).

The prospect of positive changes is a challenge to institutional leaders. The onus falls on them to coordinate the strategies that will create the desired working environment for the academic profession, and consequently improve academic quality to internationally acceptable standards.

CONCLUSION

Higher education in Nigeria is on the verge of significant changes. Once a highly regulated system, higher education is moving toward a deregulated system where government interference will be minimized. Universities in particular will enjoy administrative and financial autonomy. If properly handled, autonomy will, among other things, help counteract the loss of morale among academic staff that was responsible for the exodus of senior academics from the universities and stem the tide of brain drain. With improved funding, salaries and nonsalary conditions of service for academic staff will be further enhanced. In fact, after years of crisis over parity and nonparity of salaries among university workers an agreement has been reached between government and the academic staff unions of universities on a new separate salary scale for academics known as University Academic Staff Salary. With these developments, universities should be able to remedy the observed deficiencies caused by decades of degradation and attract academics back into the system. With autonomy, it is expected that academics will have better conditions of service to enable them to carry out their professional duties of teaching, research, and public service. Perhaps thereby the lost glory of the academic profession will be restored.

NOTES

1. From the results of the Nigerian survey, which was undertaken within the framework of the Carnegie Foundation's International Survey on the Academic Profession (Altbach, 1996), 70 percent of the respondents indicated that they had not received research grants in the previous three years.
2. In Nigeria, higher education institutions comprise universities, polytechnics, and colleges of education. The universities are degree-awarding institutions for the training of high-level manpower. Currently, there are 24 federal universities, 13 state universities, 3 private universities, and 1 military university. In addition, there are 45 polytechnics (including 21 state owned and 2 privately owned); and 62 colleges of education (including 37 state owned and 2 privately owned). These are non degree-awarding institutions for the training of middle-level manpower.
3. Estimations in U.S. dollars are based on an exchange rate of 125 naira (N125) to one U.S. dollar (U.S.$1).
4. The survey also showed that most academic staff have taught in only one or two institutions (86 percent).

REFERENCES

Academic Staff Union of Universities. (2000, August 7). What Nigerian academics want. *Punch.*

Altbach, Philip G. (Ed.). (1996). *The international academic profession: Portraits of fourteen countries.* Princeton, NJ: Carnegie Foundation for the Advancement of Teaching.

CODESRIA (Council for the Development of Social Science Research in Africa). (1996). Communiqué of the Conference on Academic Freedom in Nigeria. *CODESRIA Bulletin,* no. 1, 4–9.

Dabalen, A., Oni, B., and Adekola, O. A. (2000). *Labor market prospects for university graduates in Nigeria.* Washington, DC: World Bank.

Davies, J. L. (1985). The agendas for university management in the next decade. In G. Lockwood and J. Davies, *Universities: The management challenge.* Windsor, U.K.: NFER-Nelson.

Fafunwa, A. B. (1971). *A history of Nigerian higher education.* Lagos, Nigeria: Macmillan.

Federal Ministry of Education. (1989). *The development of education 1986–1989: National report of the Nigeria National Commission for UNESCO.* Lagos, Nigeria: Federal Ministry of Education.

Higher education in the nineties and beyond. (1991). Report of the Commission on the Review of Higher Education in Nigeria. Lagos, Nigeria: Federal Government Press.

Ipaye, B. (2000). *Quality management in the context of autonomy.* Paper presented at the National Universities Commission Training Program for Senior University Managers, Enugu, Nigeria.

Jibril, M. (2000). *Systemic overview and the challenges of autonomy.* Paper presented at the National Universities Commission Training Program for Senior University Managers, Enugu, Nigeria.

National Universities Commission. (1998). *Research bulletin* (5th ed.). Abuja, Nigeria: ADPROMO Press.

———. (2000). *1999 annual report.* Abuja, Nigeria: ADPROMO Press.

———. (Forthcoming). *Statistical digest on Nigerian universities, 1992–2000.* Abuja, Nigeria: ADPROMO Press.

Oduleye, S. O. (1985). Decline in Nigerian universities. *Higher Education* 14(1), 17–40.

Report of the Presidential Commission on the salary and conditions of service of university staff. (1981). Lagos, Nigeria: National Assembly Press.

Saint, W. S. (1992). *Universities in Africa: Strategies for stabilization and revitalization.* Washington, DC: World Bank.

Yesufu, T. M. (1973). *Creating the African university: Emerging issues in the 1970s.* Ibadan, Nigeria: Oxford University Press.

CHALLENGES AND PRESSURES FACING THE ACADEMIC PROFESSION IN SOUTH AFRICA

Charlton Koen

Much was expected to change in the academic profession in South Africa following the election of a black government in 1994. Staff equity profiles suggested that the number of black and women academics would increase significantly. Everyone expected that black institutions would receive redress funding to compensate them for decades of underdevelopment. Academics anticipated that salary levels and working conditions would improve. Others expected improvements in research output. In addressing these issues, this chapter examines a number of questions. What is the state of the academic profession in the most developed country in Africa? What terms govern work and employment conditions of academics in South Africa? What changes have occurred in the employment of black and female academics following the election of a democratically elected black majority government in 1994? How do salaries of South African academics compare internationally and what key changes are shaping the development of work practices in South Africa?

THE SHAPE OF THE HIGHER EDUCATION SYSTEM

As in many other countries, higher education institutions in South Africa operate within a regulated framework that is currently founded upon a binary structure. This binary divide is an effect of the transference of tertiary institutions and programs in education, nursing, and agriculture to higher education institutions.[1] Overall, South Africa presently has 36 public higher education institutions: 21 universities and 15 technikons.[2] The role of these institutional types has been described as follows:

> The universities and technikons are intended to be complementary sectors with formally equal status but with differentiated missions. The binary distinction between the two sectors is based on the universities' role in general formative and professional education and basic and applied research, and the technikons' role in vocational and career education and "product related" research and development. (NCHE, 1996)

Taken together, 33 are residential institutions, and 3 (2 universities and 1 technikon) function mainly as open distance-learning centers. These institutions are largely subdivided in terms of their racial origins and are either described as historically white universities (HWUs) and historically white technikons (HWTs) or as historically black universities (HBUs) and historically black technikons (HBTs). Among the universities, 11 are historically white institutions and 10 historically black institutions. Among the technikons, 8 are historically white institutions and 7 are historically black institutions.

Among the universities, language provides a further key historic marker, as some white universities were established as English-medium institutions and others as Afrikaans-medium institutions. Besides language differences, these institutions historically differed sharply by dint of their missions. First, English universities were dedicated to the intellectual pursuit of truth, justice, academic freedom, and autonomy, while Afrikaans universities were required to promote social order and to facilitate the advancement of Afrikaans speakers and the Afrikaans community. Second, in contrast to the Afrikaans universities, which did not admit any black students until the 1980s, the open English universities admitted small numbers of black students from the 1920s onward, but maintained segregation practices until the early 1970s by either excluding blacks from residences or admitting them to separate residences.

By contrast, black institutions, mainly established to promote the self-development of blacks in ethnic states, are divided in terms of ethnic classification and regional location. The origins of these institutions can principally be traced to the passage in 1959 of the Universities Extension Act. This act established the university colleges of the north (for Sotho, Venda, and Tsonga speakers), Fort Hare (for Nguni/ Xhosa speakers), Zululand (for Zulu and Swazi speakers), Western Cape (for "coloureds"), and Durban (for Indians). Later, a further three residential black universities were established in ethnically segregated "homelands" during the 1970s and 1980s and one distance-education university for black students.

The origins of technikons are more recent. Most developed during the 1970s and 1980s from colleges for advanced technical education and are similarly divided along language, race, and regional lines. Importantly, however, while historic character still defines the operation of all the institutions discussed above, almost all have shed key parts of their historic legacy. For example, instruction at Afrikaans institutions also occurs in English, and some HWU English universities now have more black (African, coloured, and Indian) than white students. Similarly, African students outnumbered coloured students at the University of the Western Cape (formerly for coloureds only) from 1996 to 2001, while African students also outnumber Indian students at the University of Durban Westville (formerly for Indians only). As a result, racial and cultural features no longer constitute the key differences between types of higher education institutions. Instead, resource differences in terms of infrastructure, funding, and research and differences in staff qualifications, staff equity, student numbers, and student quality increasingly encapsulate institutional differences.

ACADEMIC STAFF: BASIC DATA

Staff Totals

The most pertinent feature characterizing staff positions is the inequality in the distribution of senior staff by type of institution. Most staff members are appointed on a full-time permanent basis. Overall, in 2000, about 14,000 permanent faculty members (instruction and research staff) were employed at universities and technikons (DOE, 2002a). About 10,500 (75 percent) permanent staff worked at universities, compared to 3,500 (25 percent) at technikons (See table 11.1). With reference to universities, 80 percent of permanent instruction and research staff were employed at HWUs and 20 percent at HBUs. Regarding other institutional differences among permanent staff, about 96 percent of professors were employed at universities, 79 percent of associate professors, and 79 percent of senior lecturers, while most junior staff worked at technikons (CHE, 2001). Among these staff, about 82 percent of professors in the university system were employed at HWUs, compared to 18 percent at HBUs.

Given this discrepancy in senior-level staff, HWUs have stronger academic and research reputations than other institutions (NWG, 2002). Most staff also attach considerable academic prestige to employment at

Table 11.1   Distribution of university and technikon staff, 2000

	Universities		Technikons	
	HWUs	HBUs	HWTs	HBTs
Number of institutions	11	10	8	7
Number of staff	7,689	2,592	2,229	1,209
Percentage of staff	56	19	16	9
Percentage of professors	80	15	4	1
Percentage of associate professors	60	20	15	5
Percentage of senior lecturers	57	23	13	7

Note: Data from Department of Education, 2002a.

an HWU since the large HWUs are viewed as quality institutions and have good infrastructure. As a result, they are always likely to attract the better-qualified staff. However, for some staff, HWUs lack legitimacy due to their large number of white staff, conservative academic orientations, and historical legacy. This judgment applies particularly to Afrikaans HWUs and HWTs, where language and political orientations complicate interaction between black and white staff.

Gender and Race Patterns

Portraits of staff indicate that significant gender and race disparities exist and that women and blacks are overrepresented in lower-ranked jobs, despite attempts to promote staff equity (Subotzky, 1998, 2001; Cooper and Subotzky, 2001). Collectively, these portraits show that whereas in 2000 women comprised 52 percent of the national population, they constituted 38 percent of academics at both universities and technikons. (See table 11.2.) Looking at the positions women hold further reveals the extent of disparities in rank. This shows that women still remain underrepresented at senior levels, as 13 percent of professors, 24 percent of associate professors, 36 percent of senior lecturers, and 49 percent of lecturers were female in 2000.

In each equity group, the proportion of women and of Africans, coloureds, and Indians has changed only slightly since the early 1990s, with the number of white male staff correspondingly decreasing. For example, whites constituted 83 percent of permanent instruction and research staff in 1986, 76 percent in 1992, and 73 percent currently, while African staff increased from 11 percent in 1992 to 16 percent in 2000 (Bunting, 1994; DOE, 2002c). Similarly, women increased from 33 percent in 1992 to 38 percent in 2000,

Table 11.2   Gender patterns, 2000

	Male (%)	Female (%)
National population	48	52
Staff in 2000	62	38
Professors	87	13
Associate professors	76	24
Senior lecturers	64	36
Lecturers	51	49

Note: Data from CHE, 2001.

due largely to the replacement of white male staff. Nonetheless, policy documents continue to highlight the need for more significant staffing changes and underscore the point that white and male staff replacement occurs at a slow pace (CHE, 2001; NWG, 2002). This slow change in staff replacement particularly applies to Afrikaans HWUs and HWTs since the number of black academics at these HWUs remains minimal. Of further interest is the fact that white women comprise close to 70 percent of female employees and have been the main beneficiaries of equity changes.

Qualifications

The benchmark for most permanent appointments at universities remains the Ph.D. The demand is also great, since the number of Ph.D. holders is low. As can be seen from table 11.3, in 2000, 32 percent of permanent staff at universities and technikons held a doctorate, 29 percent a master's degree or master's level diploma, and 39 percent a lesser qualification. Regarding doctorates, 36 percent were awarded to professors at universities compared with 19 percent awarded to associate professors, 28 percent to senior lecturers, and 11 percent to lecturers.

Few technikon staff have received doctorates. Indeed, national figures on qualifications show that whereas 42 percent of academics at universities have doctoral degrees, similar qualifications were awarded to only 6 percent of technikon staff. Ninety percent of these doctorates were obtained at universities, indicating that technikons rely on universities for senior-level staff. Technikons also often rely on industry for junior appointees—hence the large number of staff (41 percent) with a four-year honors-equivalent qualification.

Table 11.3   Highest, most relevant staff qualifications at universities and technikons, 2000

Degree (or equivalent)	Universities (%)	Technikons (%)	Total (%)
Doctoral degree	42	6	32
Master's degree/diploma	31	23	29
Other graduate: honors/ diploma or certificate	16	41	22
Other first bachelor's/ diploma or certificate	5	22	11
Other + unknown	5	8	6

Note: Data from Department of Education, 2002a.

Most doctorates are awarded by HWUs. In 1998, technikons contributed 2 percent to the number of doctoral graduates. Most of these qualifications were awarded in science, engineering, and technology, where research and student funding levels are highest and where strong institutional support for students exists. The humanities have for a long time graduated the most Ph.D.s, but this is changing as funding levels and institutional support for staff development is expanding. In combination, these changes could positively affect lecturer recruitment in the future, as some staff development programs link recruitment and training of junior staff to enrollment in Ph.D. programs. On the other hand, many institutions support staff development programs with great reluctance, as they fear poaching and staff loss and cannot afford to invest substantial sums in such programs.

Age Levels

In terms of age, academics are fairly old, considering that average life expectancy in South Africa is about 60 years. In 2000, 23 percent were under 35 years, 63 percent were aged between 35 and 54 years, and 14 percent were older than 55 years (the age at which voluntary retirement is currently available to staff at institutions concerned with cutting salary costs). Age levels vary considerably by rank. Among professors, in 2000, 33 percent were 55 years and older, compared to 18 percent among associate professors, 13 percent among senior lecturers, and 5 percent among lecturers. Conversely, 48 percent of professors were aged between 45 and 54 years, compared to 44 percent of associate professors, 35 percent of senior lecturers, and 20 percent of lecturers. Collectively, this indicates considerable overlap among seniority, qualifications, and age, since lecturers are clearly on average

Table 11.4   Age levels by rank at universities and technikons (percentages)

Age levels	Professor	Associate professor	Senior lecturer	Lecturer	Other	Total
Under 25 years				1	14	2
25–34	1	4	15	36	45	23
35–44	17	33	37	38	22	33
45–54	48	44	35	20	15	30
55–65	33	18	13	5	4	12
Over 66	1					1
Total	100	100	100	100	100	100

Note: Data from National Department of Education, 2002c.

younger than other groups and are less qualified, while professors are on average older and hold the highest qualifications. (See table 11.4.)

EMPLOYMENT CONDITIONS

Type of Duties

Universities and technikons assume responsibility for employing academic staff in tenured positions in disciplines and departments. Generally, the department head assigns staff duties. In some departments teaching workloads are distributed through collective decision making. These loads include unequally distributed administrative work. Generally, the extent of this inequality relates to duties associated with internal positions such as course coordinator or member of a postgraduate committee. Beyond this, all permanent academic staff have teaching, research, administrative, and service responsibilities. This is shared with an increasing number of contract temporary staff, although contract staff members are not necessarily expected to undertake research. However, whereas few contract staff at technikons do research, most do research at universities in order to improve their qualifications, position, and status.

Job Types and Requirements

Regarding job types, academic job classifications or ranks of permanent staff at both universities and technikons range from below junior lecturer (including senior laboratory assistants) to junior lecturer, lecturer, senior lecturer, associate professor or associate director, and professor or director. Generally, these ranks denote teaching functions, although academics are not appointed on the strength of teaching records. At the lower levels, qualification level remains the key criterion for appointment due to the limited number of academics with doctorates and low national publication rates. For the junior lecturer and lecturer categories at universities, a master's degree is generally required, although doctorates are sought after at universities. At the upper level at universities, significance is increasingly being assigned to possession of a doctorate from a reputable institution and some accredited journal publications, while evidence of scholarship, teaching performance, and the extent of committee work feature strongly in promotions. At technikons, a master's degree is currently required for appointment as a senior lecturer.

Tenure

Employment takes two forms—contract work and permanent employment. No tenure track as in the United States exists. Instead, permanent employment typically involves a fixed-term appointment that extends to retirement. Upon appointment, the normal probationary period is 18 months. This period is sometimes extended to enable staff to be appointed to permanent positions, but permanent appointment is rarely denied. Dismissals are also rare since continued employment is guaranteed after completion of probation periods.

However, the degree of job security is changing. Dismissals have lately taken place in isolated cases. First, in these cases, retrenchments have taken place particularly at HBUs to reduce expenditure on salaries and to close departments that draw few students. Linked to this, voluntary retirement packages were recently offered to staff aged 55 years and older, to reduce numbers. Outside of this, retirement is not compulsory. However, retirement age is often linked to statutory state obligations. This prescribes that women are eligible to draw state pensions from the age of 60 years. For men, the state pension age is 65 years. Institutions do not necessarily adhere to these norms. At some, the retirement age is 60 years; at others it is 65 years. Some institutions also have early retirement policies to accelerate the appointment of people from designated equity groups. Second, management conflict at three institutions involving senior academics has been moderated through negotiated dismissals and payoffs. This practice of paying off staff for years of service when conflicts arise is not uncommon and is sometimes initiated by voluntary staff departures.

Contract Work

As with permanent full-time staff, temporary employees are not homogenous, but differ by rank, role, type of contract, length of service, and highest qualification. The incidence of temporary appointments is increasing due to the low salary costs attached to such positions. At some institutions headcount figures show that temporary appointments account for 50 to 60 percent of academic staff—up from the average of around 20 percent that prevailed at many institutions during the period between 1990 and 1995. Males comprise about 60 percent of temporary instruction and research staff at both universities and technikons. White males, at about 55 percent, are the largest numerical group among males. Among females, white women comprise close to 70 percent of temporary staff at universities because of higher qualifications, although black females are the most underrepresented group in higher education. However, of late, to meet employment equity criteria, more black males and females are also being employed.

Mostly, contract staff members have limited terms and responsibilities. Temporary employment mostly involves low-cost staff replacement strategies. It is not uncommon for senior positions at universities to remain vacant for extended periods due to the absence of suitable candidates. In some cases, replacement strategies involve the appointment of two junior staff members to fill one senior position to provide greater teaching support. In other cases, permanent staff take sabbaticals to improve qualifications, leading to the recruitment of replacements and increased teaching loads and supervision responsibilities for other staff. Beyond this, temporary appointments are common in academic student support programs and in newly established projects.

Reasons for the use of temporary staff are diverse. The increase in the number of temporary staff from the late 1980s onwards overlaps with the increase in the number of students and massification. It also corresponds with an increase in the number of courses taught to undergraduate and postgraduate students and the increasing reliance on permanent staff to teach new postgraduate courses. In these terms, the increase in temporary contract work is linked with an important status divide that points to clear teaching hierarchies. In this hierarchy, temporary staff mainly teach first-year and part-time students, with Ph.D. holders and professors mainly teaching senior students. In other cases, part-time staff teach specialist courses at senior levels. Two other distinguishing features are efforts to reduce expenditure on salaries while protecting permanent staff from excessive workload increases.

At universities, typically, temporary contract appointments of less than one year occur especially at lower levels in academic departments. This restriction (under a year) circumvents national labor relations laws that suggest that employees on one-year contracts should be treated as permanent staff if their contracts are renewed. Increasingly, longer-term, three-year contracts are also being considered to recruit staff and to make employment conditions more attractive. Other motives include laying a firmer basis for staff development, limiting turnover among temporary appointments, facilitating staff retention, and creating more stable employment conditions. Indeed, for several academic contract staff members, their working life at higher education institutions is not transitory. In some cases, a number of contract staff have been employed at the same institution for more than ten years.

At technikons, part-time staff numbers are increasing. In some departments, part-time staff outnumber permanent academic staff. Mostly, technikons recruit part-time staff from industry. Many are employed for short periods and are paid on an hourly basis. A small number are employed on one-year contracts, which are invariably renewed. Due to their background in industry, few part-time staff have teaching experience when first recruited. As a result, permanent staff in departments at some technikons often undertake capacity training that focuses on pedagogic issues. At some other technikons, instruction support centers exist to improve the quality of teaching.

In research units, institutes, and centers, contracts tend to vary from renewable one-year appointments to three-year appointments. Yet, "researcher" or "research fellow" is not an official job title at most institutions. Instead, researchers are sometimes appointed under titles such as lecturer or scientific officer. Where this happens, researchers do research full-time, whereas academics are expected to divide their time along the lines of 80 percent teaching time, 20 percent research time. Sometimes researchers also teach to improve their training portfolios and chances of securing permanent academic jobs. Titles such as lecturer or researcher and senior lecturer or senior researcher are seen as equivalent. Other positions include director, research fellow, research associate, and research intern (generally master's graduate students supported by grant funding).

UNIONIZATION, INCOME LEVELS, AND SOCIAL MOBILITY

Staff Associations

Staff negotiation with management largely centers on salaries. No functioning national union exists at universities, although the idea is strongly rooted. The origins of this idea can be traced to resistance of academic staff to apartheid policies in the 1980s. This resistance took several forms involving solidarity actions with student protesters, voluntary participation in the self-imposed academic boycott of South Africa, refusal to apply for state funding to undertake research, and protests at state dismissals and imprisonment of some employees. Institutionally, black staff members also often established new staff associations to protest the existence and operation of all white staff associations and to create space for new policy actors within institutions.

These university-based associations eventually established a national political association for progressive academics at universities in the late 1980s to coordinate antiapartheid staff activities. This structure is currently dysfunctional, as the common purpose that bonded members has dissipated and many of the initial leading figures have left universities. In its place a trade union that draws together staff from a few institutions and has little more than 1,000 members was formed (Webster and Masoetsa, 2001). As a result, local staff associations mainly undertake collective bargaining. In terms of technikons, one national staff association that negotiates salaries, benefits, and working conditions for technikon staff exists.

Most of these staff associations are weak and fragmented, which means, first, that academics lack a national voice and have recently not been a part of the national policymaking process. Second, local staff associations undertake collective bargaining. Historically, white-led racially divided associations in the past mostly accepted management salary offers while negotiating other institutional benefits. Especially at HBUs these racially established staff associations later merged with the new black associations, with staff from the latter structures playing a more prominent subsequent role due to their greater legitimacy. A second important factor responsible for their ascendancy was staff replacement strategies that largely involved the replacement of conservative white staff.

Nonetheless, fragmentation continues since staff support for associations is poor. Indeed, most associations have small executive structures and few paid-up members. Generally, executive structures tend to come together to address salary, retrenchment, and institutional benefit issues. One recent departure involved a white staff association at an Afrikaans HWU, raising concern about employment equity-related staff replacement strategies. Also, few institutions employ staff or pay for staff activities. As a result local associations are inactive for extended periods during any given year due to poor support and a lack of external pressure.

Regarding salary negotiations, some associations have refrained from negotiating increases in individual years. Increases typically involve a yearly notch increase plus a negotiated increase. Where increases are negotiated, these are often backdated as negotiations often start late. To rectify this, in some cases, academic staff associations and unions negotiating for administrative staff are considering sharing resources and negotiating joint increases. Mostly, increases correspond to the prevailing inflation rate or are set in line with institutional funding constraints, which are severe. For example, at one institution staff in some categories complain that they have not received increases for the last three years. In another case, annual salary increases are expected to be less than 5 percent for the subsequent five-year period.

Salaries

Salaries at public higher education institutions are low in comparison to what South Africans with similar qualifications earn in the private sector. Salaries are also low when compared to those available in the public sector and at research councils and foundations. In these spheres salaries have increased significantly since the establishment of a democratic government in 1994. This has resulted in large numbers of senior staff leaving to lead research divisions at public research councils. Converted to U.S. dollars, for most of these former staff, the average salary increase involves a per annum increase of $10,000. By contrast, salary increases in higher education remain constricted due to institutional debt concerns. Other problems also mitigate significant pay raises. These include the fact that salaries for all staff consume 70 percent of expenditure at some institutions, that salaries depend on the level of state funding and that state funding is likely to be reduced.

Regarding current pay scales, across institutions pay scales differ by qualification, rank, seniority, age, past work experience, and years of service. Although limited national data exist, it is becoming clear that considerable variation exists between pay levels at some institutions. This difference relates to varying institutional capacity and quality. It is also clear that financially viable institutions such as HWUs pay higher salaries. As a result, black staff, due to their concentration at HBUs, earn less than white staff members who are mainly employed at HWUs. At HBUs and HBTs recruitment is generally done in relation to existing pay scales due to institutional constraints. In some cases, more so at HWUs, individuals negotiate salary scales and benefits.

This particularly applies to senior-level appointees such as professors and deans. Broadly, salary bands for permanent employees, presented in notches, range from $7,800 per annum at the bottom end to $22,500 at the upper end at some institutions. Including associated benefits such as housing allowance, medical aid, pension, group life insurance, and annual bonus payments, the gross remuneration range changes to $12,200 at the lower end and to $32,000 at the upper end. Higher salaries are also paid to A-rated researchers who draw large research funds, have valuable networks, and constitute important assets at HWUs where they are concentrated.

In the case of researchers, salaries are generally equivalent to those of permanent staff or slightly higher. Salaries are sometimes slightly higher, often in an effort to compensate part-time research staff for the absence of other benefits such as pension, medical aid, or housing allowance. In the case of contract teaching staff, salaries tend to be lower than those paid to permanent staff enjoying similar ranks where salaries are drawn from fixed faculty budgets. This depends on the size of the available budget and the competing demands made by other departments. These factors sometimes lead to some contract staff being paid at nonnegotiable rates that vary from $14 to $30 per hour. Beyond this, salaries of part-time lecturers are mostly nonnegotiable and tend to vary from $300 to $700 per month.

Do academics form part of the middle class in South Africa? Yes, salary levels and remuneration packages place them among this economic interest group. However, while salaries of middle-income earners have increased substantially of late, the scale of these increases did not extend to academics, unless they undertook private contract work. Do academics lead comfortable middle class lives? No easy answer exists. For many, the standard of living and quality of life are good, although sharply rising food prices and significant increases in private medical tariffs and in car and house prices have occurred lately. The average cost of living is also high. In 1999 interest rates peaked at 24 percent, after averaging between 12 and 14 percent in the early 1990s. Currently, the level is 15 percent, but this seems set to increase over the next year. Also down from average levels of 11 percent for the period from 1998 to 2001 is inflation. This is currently 6.1 percent, but also seems set to rise and to erode disposable income levels.

In these terms, single academics at the lower end of salary scales and even in the middle will find great difficulty in paying off both a house and a car. Few at the lower end could do this and invest in personal retirement savings or other financial security schemes. Salaries at the lower end are also not attractive, as consumer debt burdens are high and many new potential staff comprise the first generation of middle-income earners in families. Some of these individuals are ultimately responsible for family and household income and often need to share resources with younger siblings and other family members. Therefore, some experience great pressure to earn high salaries and to forsake academic employment since their skills also equip them for other jobs.

Other Income Sources

In terms of other income sources, specific institutional benefits are also available—such as bonus payments to department heads. They also benefit from lighter teaching loads, while faculty deans, deputy deans, and other management-level staff members earn substantially more and control research money and entertainment allowance funds. Nominal teaching bonuses are also paid. In some cases, travel bonuses are also paid, but not necessarily as a perk. This practice is especially prevalent at HBUs situated in underdeveloped rural areas that depend on staff from large towns. In these cases staff members are compensated for their long commutes to work.

Staff members also often supplement their income through consultancy work. Within institutions clear expectations exist around consultancy work. These expectations include doing limited consultancy work, informing institutional structures about additional income sources and gaining permission for extra work. Rules forbid consultancy work at some universities, but the practice is also extensive at these institutions. At other institutions, regulations govern the size of additional income sources and the impact on teaching and allow for institutions to be paid compensation fees. However, regulations are often flouted as direct controls do not exist, effective monitoring is well nigh impossible, and many recognize that academic salaries are low in comparative terms. Concomitantly, no good data are available regarding the extent of extra income sources and their level.

Staff Mobility

Typically, department structures at universities contain one professor, an associate professor, an elected department head, senior lecturers, lecturers, junior lecturers, and administrative assistants. More complex structures also exist in interdisciplinary schools and programs that draw together two or more departments. In other cases, some departments, especially at HWUs, are top-heavy with senior staff due to promotions. For example, one university-based education department is comprised of two lecturers, seven senior lecturers, and other more senior staff. This is a result of the recent introduction of greater flexibility into post structures to facilitate the advancement of underrepresented groups. A secondary factor is retaining staff by providing for accelerated promotion. In form, this is associated with the development of a career track in which promotion is linked to performance and reputation. This has replaced the former restricted promotion system, in which advancement depended on the creation of new posts or the replacement of more senior staff.

At technikons, the conventional department until 1998 consisted of lecturers, senior lecturers, a department head, associate directors, and directors. Typically, promotions depended on qualifications, teaching portfolios, and years of experience. For example, promotion from lecturer to senior lecturer depended on acquisition of a master's degree and two years of teaching experience. This is changing since newly appointed lecturers increasingly require a master's degree qualification. The typical department structure is also changing. Developments over the last three to four years have resulted in the appointment of associate professors and professors. Along with this, a new career track linked to research output, industry links, and publications—rather than qualifications, teaching, and administrative work—is opening up. Initially, career track routes involved promotion from lecturer to senior lecturer, head of department, and director. Now a senior lecturer can also become an associate professor or professor. One further effect is the addition of an academic leader to complement the work of department heads who function as managers. However, while mobility opportunities are increasing at one level, they are constricted at others due to extreme financial stringency's and internal restructuring. Together, these factors have resulted in promotions being put on hold at some technikons.

By contrast, mobility opportunities at universities are diverse. Responsible factors include expanding opportunity structures both inside and outside higher education and policy support for employment equity. One measure of movement can be obtained from the loss of senior staff with established publication track records. Much of this movement has involved the departure of midlevel and senior staff from HBUs to HWUs, technikons, and state-funded research organizations. Some of these individuals moved to more prestigious universities to gain access to increased funding opportunities and better working conditions. Others moved to technikons and research bodies that offer senior status, substantially higher salaries, resource-intensive environments, access to new networks, and the opportunity to escape from teaching responsibilities. For these institutions such recruitment is essential to improve their staff equity profiles, increase their own legitimacy, promote change within their own institutions, and bring in individuals that can help transform HWUs.

Not all of this movement has involved moves to other teaching positions. A central changing feature of the higher education landscape involves the expansion of senior-level management positions and the improvement of institutional supports for students and senior managers. This opened a new managerial track for academic staff that filters into senior administrative positions where knowledge of academic work is important. A second feature in academic departments at HBUs involves junior faculty replacing senior staff, with employment equity promotions being offered to internal and external candidates. Currently, accelerated promotion is also common to associate professor and professor positions to fill vacancies and to reward white and black staff for long service. In addition, several institutions have set equity targets at senior levels and aim to fill these positions through peer-nominated promotion, while career tracks linked to assessing teaching portfolios, community service work, committee work, and research are opening up. In this sense, recent faculty changes induced greater flexibility and mobility into previously rigid academic structures and helped facilitate the upward mobility especially of staff with recently obtained Ph.D.s.

In research units (almost exclusively found at universities), progression is sometimes rapid. In most cases, research units are small with less than six staff members. Most experience rapid turnover in staff since employment conditions are impermanent and research staff tend to be studying toward higher qualifications. As a result, few units invest in training of research interns, research capacity building programs, or long-term projects. Progression to the rank of senior researcher is also often quick since staff tend to depart to pursue more permanent, but less prestigious, careers in teaching posts and because vacancies are often filled internally. By contrast, larger research units have a more fixed post structure, work in defined areas, show greater staff stability, and follow department norms regarding promotion.

WORKING CONDITIONS

General Features

Working conditions vary in relation to the number of students and resource constraints. Generally, due to massification and budget constraints, workloads are intensifying. At universities and technikons, staff to student ratios differ by field of study. At some institutions 1 : 25 is viewed as the norm in science, engineering, and technology and 1 : 35 as the norm in the arts and humanities. In practice, staff-student ratios are appreciably higher in specific fields as figures of 1 : 70 are bandied about. This variance relates to the uneven impact of massification. This has recently induced sharp increases in student numbers at technikons. Massification was first experienced at universities with staff–student ratios at HBUs in disciplines such as sociology, in one case, of about 1 : 200 during the early 1990s. Since then, much has changed due to the effect of student debt, poor retention levels, and dropping enrollments. This has produced substantially improved current ratios of 1 : 50 in disciplines where much higher ratios were previously recorded. Along with this, working conditions at most HBUs have recently improved significantly, after deteriorating appreciably in the early 1990s.

Typically, the academic year spans 28 to 32 teaching weeks. This is divided into two semesters, which are subdivided into four quarters separated by a ten-day spring break, four-week winter vacation, ten-day autumn break, and ten-week summer break. Staff responsibilities include teaching, marking, supervising, consultation, committee work, and research and service activities. Regarding other expectations, at universities academic staff are typically expected to spend the work week at institutions and are given one day off for research each week. In addition, they are expected to do committee work and teaching over 4 days and can take 21 days leave (not including medical leave), but are expected to work a full day and to be available for consultation and supervision. A further general benefit relates to study or research leave. As in some other countries, this involves one full year's leave for every six years of work. Staff members are further expected to undertake course evaluations, but do this infrequently, although it is described as compulsory in policy documents at some institutions.

Teaching and Marking

For most academics, teaching and academic development or support remain the prime responsibility. At universities, teaching loads are minimal for most staff. Ogude, Mavundla, and Netswera (2001) report that the average teaching load of a lecturer at universities is 6 to 8 contact hours per week. This varies substantially across HWUs and HBUs and between faculties. For example, workloads of academics at HBUs in sociology involve teaching two to three courses per year, compared to three to four courses per year at HWUs in sociology.

Besides teaching load, the number of contact hours also differs. Thus, in sociology at one HBU a lecturer spends 1 to 3 hours per week in a class, whereas the counterpart at an HWU spends 3 to 6 hours. Contact time in science, engineering, and technology programs at both HBUs and HWUs are also greater due to the combination of lectures, tutorials, and laboratory work. This averages 6 to 12 hours per week, with senior staff on average having less teaching responsibilities than junior staff.

At technikons, in some programs estimated teaching-related activities range from 12 to 24 hours per week. This often varies by rank, with department heads and senior staff having less contact time. Sometimes the nature of this division relates to the amount of state subsidy money staff members generate, rather than staff–student ratios. One result is that junior staff members do more teaching to larger classes at lower levels with senior staff taking smaller classes and more research students at the upper levels. In addition, lecturers are expected to have at least one compulsory consultation period per week, but are also available during other times at some institutions in faculties where core time systems are used. This core time system involves an expectation that, over and above other teaching responsibilities, lecturers should physically be present at the technikon during set hours each day. This system is largely a response to changes in management personnel and efforts to improve productivity levels.

Generally, marking loads are limited in science, engineering, and business and commerce fields, but are intensive in arts and humanities courses at universities. This sometimes involves marking assignments of more than 100 underprepared students at a time. Student development problems include poor command of English (the main language of instruction at 19 universities and at most technikons). For most students, English is a second or third language. Indeed, national census statistics indicate that English is a first language for fewer than 20 percent of the population. Many students also accordingly enroll for English development courses during their first year of studies since this is a compulsory part of academic development at some institutions.

Beyond these features, the characteristics of academic work are not well documented. However, some features are changing. At some institutions pressure to increase income from higher student numbers has resulted in academics becoming involved in marketing and recruitment drives and undertaking mentoring activities. At most institutions greater effort is also being devoted to increasing involvement of academics in committee work. This is crucial since some staff at technikons have no committee responsibilities. In this sense, it is clear that expectations about the scope of academic work are increasing. For example, staff members are increasingly expected to produce one accredited journal publication per year, although, no prescriptions exist.

Research Output

Regarding research output, productivity increases have occurred of late as institutional reports to the national Department of Education show a steady overall growth of around 25 percent in research output from 1986 to 1996 (Bawa, 2001). This increase flowed mainly from a sharp increase in research output from HBUs from 1986 onward that resulted in their total share of research productivity increasing from 5 percent in 1986 to 12 percent in 1998, after which it decreased to 10 percent in 1999 (CHE, 2001). More specifically, research productivity increased sharply over this period at two specific HBUs—leading to their academic reputations improving significantly.

In line with earlier institutional descriptions, research performance remains uneven.[3] In 1999, HWUs were responsible for 86 percent of research output, HBUs for 10 percent, with technikons accounting for 4 percent (Bawa, 2001). More significantly, the top 6 universities (out of 21) contributed close to 70 percent of research publications cited by the *Science Citation Index*. These top 6 institutions in terms of research output are all HWUs. Other key features include their location in the three most powerful economic regions of the country and, in comparison to other institutions, their strong concentration of research professors and selective student admission policies. In the case of technikons, the 5 best performing in terms of research output are also HWUs (Ogude, Mavundla, and Netswera, 2001) and are concentrated in the same economic regions as the best performing research universities.

Overall, research output is roughly in line with research and development expenditure (including staff time) at higher education institutions. Diverse sources indicate that technikons continue to spend less than 5 percent of the research funding in the higher education sector. HBUs spend 10 percent and HWUs spend 85 percent. In total, publications numbered slightly more than 5,000 units in both 1998 and 1999 (Bawa, 2001)—indicating that less than 50 percent of permanent staff published. Publication output is also a problem at research units. Research units are often prolific in producing reports, but seldom convert these outputs into accredited publications.

Reasons for the low publication rates are diverse, but can be linked to what appears to be intractable problems.

First, the training of academic staff and researchers remains limited and poor. In particular, many lack basic skills in methodology and statistical analysis. Second, research funding organizations continue to express concern about the small number of applications in some fields. Third, several accredited journals appear infrequently and urgently need substantial funding to maintain their long-term viability. Fourth, some staff remain reluctant to undertake research and instead view teaching as a vocation. Fifth, outside of some pockets of research enterprise, strong research and academic cultures are not enduring features at HBUs and technikons. Sixth, academic leadership in departments and in research units remains weak, and library holdings and the quality of library service continue to need substantial improvement at most institutions.

SOME MAJOR DEVELOPMENTS AND RECENT TRENDS
Brain Drain

Higher education in South Africa is at crossroads. As can be detected from the preceding section, several challenges exist. These include the problem of brain drain and the need to upgrade qualifications. In each of these cases, it is difficult to gauge the extent of the problems higher education faces. The brain drain phenomenon is real. Unconventionally, the main source is not the limited loss of staff to teaching institutions (mainly) in Australia and the United Kingdom. This has chiefly involved the loss of small numbers of senior white male staff in social science, commerce and financial sector fields to countries that traditionally have attracted English- and Afrikaans-speaking whites from South Africa. However, the key feature is the departure of senior male colleagues from mainly three HBUs to higher-paying jobs in the public sector and the movement of black staff from less- to more-prestigious institutions. What distinguishes these three institutions from other HBUs is largely the greater strength of their academic programs and past strong political support for democratic rule in South Africa.

A further distinctive feature is the age profile of those white and black males who leave. In line with levels of seniority, departing academics fall mainly in the 40-to-55-year age bracket. The lower end of this age bracket typically includes those who have recently obtained a Ph.D. The upper end is typically associated with academic leadership and an established publication record. In this sense, the staff loss is significant, as indications of declining research output is already evident and relatively inexperienced leaders head several departments. In addition, replacement is bedeviled by the poor qualification profile of permanent and contract staff and institutional debt concerns.

However, while staff losses at HBUs have occurred mainly in education and social science fields, these departures have been offset by several factors. For example, in many ways, this movement constitutes a benefit to the higher education system as many erstwhile colleagues work in closely related sectors and channel research and other development funds to higher education institutions. Also, some level of staff stability is also evident, while senior staff is increasingly being recruited from other African and European countries. For example, figures for 2000 indicate that foreign nationals constitute about 5 percent of permanent instruction and research staff (DOE, 2002d). Demographic features indicate that about 75 percent of these staff members are male, 40 percent come from an African country other than South Africa, and close to 40 percent (of whom more than half are employed at two English HWUs) come from a European country. In this sense, resource depletion has been augmented by the infusion of new staff and greater demographic and intellectual diversity. For many staff and students this is a rewarding development.

Mainly, this change in the recruiting base of staff is a consequence of the change in government from the apartheid oligarchy to a democratic order. This lifted the self-imposed academic boycott that existed from the mid-1980s to the early 1990s and contributed to several institutions' deciding to embrace internationalization practices. Other consequences of this changing environment include more regular visits from overseas academics; international research cooperation agreements and projects; the greater infusion of universal perspectives into higher education; and more regular exposure of South African academics to international academics and conferences. In this sense, systemwide, the changing academic environment is benefiting immensely from a stronger emphasis on internationalization practices.

On the other hand, the effect of brain drain is starkly visible in departments. For example, in some departments senior staff are comprised solely of individuals who obtained Ph.D.s in the last three years. In many other departments, promotions are tackled hastily to fill gaps left by departing senior staff. Sometimes such promotions merely reward staff for years of service or for newly acquired Ph.D.s, with little significance being attached to the quality of the degree. Such practices could have damaging long-term effects, as brain drain

has particularly increased the supervision load of newly graduated
Ph.D. holders and inexperienced staff. The same conclusion applies to
employment equity promotions. By law, all institutions need to devise
employment equity plans and set targets. Tensions exist between
reliance on merit principles and the use of equity criteria in recruiting
and in promotions since both types of criteria are necessary.

Beyond this it seems important to recognize that brain drain signals that staff are highly mobile. It is also important to recognize that
brain drain is interrelated with factors such as low remuneration levels, increased staff workloads, institutional instability caused by student debt, intellectually unexciting environments, new career
opportunities, and low staff morale. This has not been measured but
is often mentioned. Yet, while staff members depart, few institutions
have staff retention or staff development plans and strategies. Instead,
ad hoc responses proliferate when resignations loom. At times, this
involves counteroffers of promotion and better pay, suggesting that
greater attention should be paid to academic career paths. A further
strategy involves seconding staff to external research centers in order
to delay departures and to stabilize employment patterns.

Of course, these management responses at some institutions also
need to be understood in the context of fiscal constraints and the need
to free positions in order to effect equity criteria. A further variable
involves increased competition between institutions to attract senior
staff and to improve chances of accessing research funds. As indicated
earlier, HBUs, in particular, experience great difficulty in competing
with HWUs. For example, in one case, an HWU was able to recruit a
staff member to a vacant lecturing position, although the applicant was
offered a senior lecturer position at an HBU. The key factor influenc-
ing the decision was the offer of a higher salary and better bonuses. Of
some historic irony here is the fact that material factors drive these
recruitment initiatives, whereas staff recruitment during the late 1980s
was partly linked to the need to reinvigorate intellectual environments
and to draw together a critical mass of intellectuals.

Academic Autonomy and Institutional Decision Making

Overall, academic autonomy can be understood in two ways: freedom
to determine what is taught, to express views, and to operate without
external state interference (Du Toit, 2001). Today, academic autonomy remains contentious. Culturally, academics are free to determine
the content of curricula. However, the recent establishment of new
quality assurance structures and the expectation that content should

be linked to outcomes will likely introduce greater uniformity in programs across institutions. More broadly, academic freedom to express views is not unfettered. In one recent case, a staff member was dismissed for criticizing management. In other instances, staff members have been asked to explain their published reflections on academic experiences—indicating increasing sensitivity to criticism within institutions. Much of this sensitivity relates to increased competition between institutions and the recognition that competitors exploit criticism.

Regarding state interference, all institutions enjoy autonomy. However, recently government passed the higher education amendment act to allow the minister of education to determine the purpose and function of institutions. This allows for prescriptive policies and effectively erodes institutional autonomy. Collectively, government influence derives from several sources: laws, provision of subsidy funds, policy formulation and implementation, and ministerial appointees to university and technikon councils. In addition, stipulations in the act allow the minister of education to appoint temporary managers at higher education institutions where leadership crises exist. A further recent intervention resulted in a court case questioning the extent of the minister's powers following initiatives to merge a distance education university and its technikon counterpart. This case called into question the mediating political role government sometimes plays in policy issues.

This role is particularly evident from recent student boycotts, during which government officials met with student leaders following the breakdown of their negotiations with management. Such meetings have not focused solely on reestablishing lines of communication between students and institutional leaders. Instead, in one case settlement terms were offered and leaders of one national student organization, sympathetic to government policies, stepped in to persuade local student leaders to end their boycott. In this way, higher education is not free from government interference, but academics do control their own activities and function within the ethos of a prevailing culture that emphasizes academic autonomy and autonomous institutional decision making.

Quality Assurance and Peer-Review Mechanisms

One significant looming change involves the extension of quality assurance mechanisms. Recently, in 2001, the Higher Education Quality Control (HEQC) framework was outlined. Together with the recent establishment, since 1997, of the South African Qualifications Authority, National Qualification Authority, National Standards Bodies, Standards Generating Bodies, and the Sector Training Authorities, the HEQC is responsible for approving the outcome of training, accrediting courses, and monitoring and reporting on institutional efficiency. In this sense, quality control mechanisms provide a platform to establish more uniform standards across institutions and to establish benchmarks for evaluating the performance of staff (CHE, 2002). Specifically, this highlights systemic efforts to link skills' development to knowledge transfer and to introduce greater accountability into teaching. For many, it also signals national efforts to assign greater weight and value to pedagogic concerns and to provide a firmer reward basis for good teaching.

Currently, quality assurance practices are limited. In disciplines such as psychology, engineering, and law professional bodies undertake periodic reviews. In other disciplines, external examination boards regulate entry into professions. Across disciplines, examinations are moderated internally at lower levels and externally at senior levels at universities and technikons. At both institutional types, staff members from other academic institutions undertake the moderation. This moderation follows strict criteria that are applied uniformly. At technikons, moderation also extends to examination reviews by staff from industry through the activities of advisory committees that exist in all disciplines and through professional bodies.

However, generally missing are systematic peer review, independent assessment, and efficiency indicators on departmental and individual staff performance with respect to teaching. Where teaching review mechanisms exist, they are more likely to be found at HWUs (where evaluations are more common) than at other institutions. Staff evaluations do, however, occur indirectly as department heads compile annual reports on research productivity and teaching loads. Academic development staff also often assist junior staff with pedagogic issues and informally evaluate staff performance. In this sense, while formal evaluations occur infrequently, broad accountability measures to determine staff performance do exist.

A further development with peer assessment implications involves the extension of a research rating system used in the sciences since 1984 to arts and humanities practitioners. In this system, the elite group of top-performing academics is tagged as A-rated scientists, followed by B-, C-, Y-, P-, and L-rated groups. Category A refers to staff that are recognized as leading international scholars by their peers. Category B refers to researchers who enjoy considerable international recognition and C to established researchers with a sustained recent record of productivity. P and Y provide peer ratings for researchers younger than 35 years of age who have obtained doctorates and are judged to have the potential to establish themselves as researchers, while L covers all researchers younger than 55 years of age who cannot be placed in another category. In each case, researchers are expected to apply for ratings. Subsequently, documentation is referred to subject-specific specialist committees that identify the names of at least six peer reviewers. The specialist committee is required to examine the quality of research outputs by the applicant over the last seven years, to judge the applicant's standing as a researcher and the quality and appropriateness of the sources in which the applicant has published. Peer reports are subsequently referred to members of assessment panels that assign ratings (NRF, 2002).

The implications of this review process are potentially enormous. Ratings in science, engineering, and technology have shown that close to 90 percent of A-rated researchers are clustered at HWUs. HWUs also have the most B- and C-rated researchers in science, engineering, and technology and use these quality indicators to recruit students and to enlist donor support. Implicitly, this shows that institutional differentiation is also increasingly being linked to research ratings and performance. Implications also exist for individuals, as A-rated scientists command more prestige, have greater access to research funding, more research students, and have fewer teaching responsibilities. More broadly, it is already evident that research awards are driven by qualifications linked to possession of a Ph.D. This is especially true in cases where South Africans undertake joint research with overseas researchers and with regard to donor criteria that link leadership of team research to possession of a Ph.D. As indicated previously, these criteria favor HWUs and imply that the existing gap between historically white and historically black institutions will increase in the future.

CONCLUSION

Structural features associated with higher education under apartheid still remain. What is of current interest is the way in which a hierarchy of academic tasks and working conditions is emerging. The example of A-rated scientists with limited teaching responsibilities and high salaries provides one illustration. Other examples are clear from the greater use within institutions of titles such as senior professor and super professor, distinctions between full-time and contract staff, and the limited involvement of senior staff in undergraduate teaching. At some technikons, it is also evident from efforts to link teaching loads to the amount of subsidy money staff generate and reward senior staff who teach senior students with lower teaching loads.

Of some interest regarding working conditions are research suggestions that staff job satisfaction is reasonably high. In some ways this is not surprising. Working conditions have certainly deteriorated in many departments since the early 1990s, but many staff have also adjusted. In other cases, working conditions, especially at HBUs, have also improved lately due to declines in student numbers, improved managerial efficiency, and the better provision of resources. Beyond this, it is not clear that staff members are overworked although complaints of high teaching loads at technikons and of marking loads at universities are frequent.

By contrast, staff morale is much more uneven as many institutions are contemplating retrenchments to reduce salary costs and staff remain uncertain about the outcome of institutional restructuring initiatives. Factors underlying these uncertainties appear important. For example, suggestions that technikons may become universities of technology in the near future have motivated concern about improving qualifications and how restructuring will affect institutional alignments and activities. At universities, concern about job security, salaries, and remuneration levels is motivated by efforts to change the proportion of students in the system. Over the last five years, this has resulted in a quarter reduction in the number of arts and humanities students, close to 50 percent growth in business and commerce student numbers, and roughly 30 percent growth in the science, engineering, and technology streams. In the arts and humanities, this trend will most likely lead to job loss, retrenchments, and more uncertainty.

And given the high concentration of students and staff at HBUs in arts and humanities programs it is likely that the impact of restructuring will be most keenly experienced here. On the other hand, Afrikaans and English HWUs and HWTs also need more black and female staff and are most likely to recruit such staff from arts and humanities programs at HBUs. In this sense, the drive to promote staff equity may well continue to lure more established black staff away from HBUs and to weaken HBUs in the future. To compensate for this, HBUs probably need to pay greater attention to staff recruitment, staff retention, working conditions, institutional rewards, and merit systems. Doing this should contribute to greater competition, increased efficiency, and positively impact research output.

NOTES

1. This process first involved a reduction in the number of tertiary institutions, followed by the incorporation of the remaining institutions and programs into higher education institutions. This process is almost complete. For example, in education, the more than 300 non-higher education tertiary institutions have been reduced to less than 10, which all await incorporation into higher education institutions.

2. The term *technikon* is essentially a noun used to refer to technology-related teaching activities and is viewed in South Africa as a substitute term for an institute of technology.

3. It bears mentioning that some concern surrounds measurement of research output since disparities exist in the list of accredited journals used in different indices. One result is that the official South African registrar on publications probably underreflects research productivity.

REFERENCES

Bawa, A. (2001, March). *Trends in South African research system.* Paper presented at the SRHE/EPU conference on Globalization and Higher Education. Cape Town, South Africa.

Bunting, I. (1994). *A legacy of inequality: Higher education in South Africa.* Cape Town, South Africa: Cape Town University Press.

Cooper, D., and G. Subotzky. (2001). *The skewed revolution: Trends in South African higher education, 1988–1998.* Cape Town, South Africa: Education Policy Unit, University of the Western Cape.

Council on Higher Education (CHE). (2001). *The state of higher education in South Africa: Annual report, 2000/2001.* Pretoria, South Africa: CHE.

———. (2002, January). *A new academic policy for programs and qualifications in higher education.* Pretoria, South Africa. Department of Education.

Department of Education (DOE). (2001a). *Annual report 2000/2001.* Pretoria, South Africa: DOE.

———. (2001b). *National plan for higher education.* Pretoria, South Africa: DOE.

———. (2002a). *Universities/technikons: Headcount of instruction/research professionals with permanent appointments according to highest most recent qualification and rank obtained for the year 2000.* Pretoria, South Africa: DOE.

———. (2002b). *Universities technikons: Headcount of personnel with permanent appointments according to personnel category, racial/ethnic identification and sex for the year 2000.* Pretoria, South Africa: DOE.

———. (2002c). *Universities/technikons: Headcount of instruction/research professionals with permanent appointment according to age, rank and sex for the year 2000.* Pretoria, South Africa: DOE.

Department of Education (DOE). (2002d). *Universities/technikons. Headcount of instruction/research professionals with temporary appointment according to nationality for the year 2000.* Pretoria, South Africa: DOE.

Du Toit, A. (2001). *Revisiting academic freedom in post-apartheid South Africa: Current issues and challenges.* Background paper for the forthcoming CHET publication, *Higher education policy, institutions and globalization: New dynamics in South Africa after 1994.* Pretoria, South Africa: Center for Higher Education Transformation.

National Commission on Higher Education (NCHE). (1996). *A framework for transformation report.* Pretoria, South Africa: Department of Education.

National Research Foundation (NRF). (2002). *The evaluation and rating of the research performance of researchers in South Africa.* Pretoria, South Africa: NRF.

National Working Group (NWG). (2002). *The restructuring of higher education in South Africa.* Pretoria, South Africa: Department of Education.

Ogude, N., Mavundla, T., and Netswera. G. (2001, March). *A Critical analysis of the status of research at South African technikons.* Paper presented at the SRHE/EPU conference on Globalization and Higher Education. Cape Town, South Africa.

Subotzky, G. (1998, July). *Career challenges for women in academia: Facts, figures and trends in higher education in South Africa.* Paper presented at the UWC Gender Theory and Practice Winter School. University of the Western Cape, Cape Town, South Africa.

———. (2001). Addressing equity and excellence in relation to employment: What prospects for transformative change in South Africa? *Equity and Excellence in Education,* 34(3), 56–69.

Webster, E., and Masoetsa, S. (2001). *At the Chalk Face: Managerialism and the changing academic workplace 1995–2001.* Background paper for the forthcoming CHET publication, *Higher education policy, institutions and globalization: New dynamics in South Africa after 1994.* Pretoria, South Africa: Center for Higher Education Transformation.

About the Authors

PHILIP G. ALTBACH is J. Donald Monan, SJ professor of higher education and director of the Center for International Higher Education at Boston College. He is editor of the *Review of Higher Education* and author of such books as *Comparative Higher Education* and *Student Politics in America*. He has edited *International Higher Education: An Encyclopedia*, *Private Prometheus: Private Higher Education and Development*, *Comparative Perspectives on the Academic Profession*, *The Academic Profession: An International Portrait*, and other books.

ELIZABETH BALBACHEVSKY is assistant professor in the Department of Political Science of the University of São Paulo, Brazil; senior researcher at the University Research Center on Higher Education; and also senior researcher at the University Center on International Relations. She has written on the Brazilian academic profession.

MONICA IYEGUMWENA BARROW is director of the Personnel Management Department of the Nigerian Universities Commission. She also serves as facilitator of the National Training Program for University Managers and is chair of the state development subcommittee of the NUC-World Bank Implementation Unit. She holds degrees from the University of North Carolina and the University of London Institute of Education. She has written on the development of private postsecondary education in Nigeria as part of the 1996 UNESCO/CHEMS joint study of private postsecondary education in four Commonwealth countries.

XIANGMING CHEN is a researcher and teacher at the Institute of Higher Education, Peking University, Beijing, China. She holds a doctorate in education from Harvard University. She has been involved with several research projects in China, sponsored by the World Bank, United Nations Development Program, the U.K. Department for International Development, and the Chinese Ministry of Education, and has written widely on higher education issues.

MARIA DA CONCEIÇÃO QUINTEIRO is a senior researcher at the University Research Center on International Relations at the University of São Paulo, Brazil.

MANUEL GIL-ANTÓN is professor of sociology in the Department of Sociology, Autonomous Metropolitan University, Azcapotalco, Mexico City, Mexico, and is a member of the National Research System of Mexico. He is coauthor of *Los ragos de la diversidad*, a study of Mexican academics, and other books on Mexican higher education.

N. JAYARAM is professor and head of the Department of Sociology at the University of Goa, India. He is editor of the *ICSSR Journal of Reviews and Abstracts: Sociology and Social Anthropology*. Among his books are *Higher Education and Status Retention, Sociology of Education in India*, and *Social Conflict*.

CHARLTON KOEN is a researcher in the Education Policy Unit at the University of the Western Cape, Belleville, South Africa. His research interests include student politics, graduate employment patterns, and time-to-degree rates.

MOLLY N. N. LEE is associate professor in education at Universiti Sains Malaysia. She teaches sociology of education and science teaching methods. She has written on higher education, teacher education, private education, science education, and educational policies. Recent publications include *The Corporatisation, Privatisation and Internalisation of Higher Education in Malaysia*; *Education and the State: Malaysia After NEP*; *Private Higher Education in Malaysia*; *The Politics of Educational Change in Malaysia: National Context and Global Influences*; and *The Impacts of Globalization on Education in Malaysia*.

SUNGHO H. LEE is vice president of Yonsei University, Seoul, Korea, where he has also served as dean of student affairs at the college of education and at the graduate school. He is a former assistant minister of higher education in the Government of the Republic of Korea and served as a member of various presidential commissions. He has published widely on Korean higher education and on curriculum issues. Among his books are *Conflict in Korean Higher Education* and *Curriculum Development in Korean Higher Education*.

CARLOS MARQUIS is dean at the Universidad de Palermo, in Buenos Aires, and a sociologist and a researcher for the Argentine Council of Technological and Scientific Research. He was formerly a member of the Mexican National Research System. He has been a tenured professor at Argentine and Mexican universities and his research and professional field is higher education. For five years he has been Executive Director of FOMEC, a higher education reform program financed by the World Bank and the Argentine Republic.

ANDRÉ ELIAS MAZAWI is head of the Sociology of Education Graduate Program, in the Department of Educational Policy and Organization, at the School of Education, Tel-Aviv University, Israel.

FIDELMA EKWUTOZIA UKEJE is deputy director of the Department of Academic Planning National Universities Commission of Nigeria, and currently head of the newly created Strategic Planning Division. She holds a doctorate in linguistics from the University of Paris III (Sorbonne) and has published in her academic field of French studies.

INDEX